Gas Migration

Gas Migration

Editor

Akash Malik

Gas Migration

Edited by **Akash Malik**

Printed in 2017

ISBN: 978-1-68117-382-5

Library of Congress Control Number: 2015941568

© 2016 by

SCITUS Academics LLC,
616, Corporate Way, Suite 2, 4766,
Valley Cottage, NY 10989

www.scitusacademics.com

Contents

Preface

Gas Migration is the book to accumulate, analyze and apply the interdisciplinary knowledge on gas migration and detail its connection to tectronic, seismic, and geologic phenomena. It combines geological, geochemical, geophysical, seismological, and petroleum engineering insights to demonstrate how gas migration and its associated phenomena can be used in earthquake and environmental geohazard identification and prediction. Gases migrating to the earth's surface provide crucial information regarding the state and evolution of the structure and tectonics of our planet. By listening to and analyzing "whispering" gases, locations of intense tectonic activity can be identified and areas of potential natural disasters delineated, such as volcanic eruptions, tsunamis, and earth-quakes. Like an experienced doctor who can determine the health conditions of a patient by listening to his breathing, a knowledgeable scientist can recognize a disorder in the earth's body by analyzing these migrating gases. The migration of gas to the surface from oil and gas formations is a problem that greatly affects those surface areas where human activity exists. Underground gas storage facilities and oil fields have demonstrated a long history of environmental gas migration problems. Experience has shown that the migration of gas to the surface creates a serious potential risk of explosion, fires, noxious odors, and carcinogenic chemical emissions.

Editor

Calculating Efficiency of Separation of Aerosol Particles from Gases in Packed Apparatuses

Anatoly G. Laptev[1, 2], Marat M. Basharov[1], Timur M. Farakhov[1, 2], and Albert R. Iskhakov[1]

[1]Department of Technology of Water and Fuel, Kazan State Power Engineering University, Kazan, Russia

[2]Engineering-Promotional Center "Inzhekhim", Kazan, Russia

ABSTRACT

A mathematical model for the process of removal of aerosols from gases in channels filled with packing is considered. Equations for transport of aerosol particles and results of their numerical solution are given. Influence of operating conditions and of design characteristics of the packed apparatus on gas purification efficiency is analyzed.

INTRODUCTION

Gaseous environments are often used in chemical and petrochemical industries and power engineering. As a rule, the gaseous environments are of a non-uniform composition and are contaminated by disperse phase admixtures causing problems through exerting negative influence on efficiency of technological processes and the process equipment operation; therefore, they must be removed from gases.

Disperse phase separation efficiency can be determined through using the value of disperse phase content in the gas before the point of entrance into the gas purification apparatus and the value after the point of exit from the apparatus via the following expression:

$$\eta = \frac{Q_i C_i - Q_f C_f}{Q_i C_i} = 1 - \frac{Q_f C_f}{Q_i C_i},$$

(1)

where C is the particle concentration (kg/m³); Q is the gas flow rate (m³/s); subscripts i and f refer to the initial and final values, respectively.

One usually adopts for apparatuses the equality $Q_i = Q_f$.

As is known, particle deposition in apparatuses can be due to several mechanisms such as gravitational, inertial, turbulent, centrifugal and other mechanisms. Most industrial apparatuses operate through utilizing several of these mechanisms by combining them together. Hence, disperse phase separation efficiency is dependent of many different parameters.

In case of several mechanisms acting simultaneously or in case of purification of gases being carried out in series in back-to-back apparatuses (zones), the overall efficiency is calculated through using the additivity rule in the following manner:

$$\eta_\Sigma = 1 - \prod_k (1 - \eta_k),$$

(2)

Where η_k is the separation mechanism due to the kth mechanism in the kth apparatus (zone).

In the present, work we use the approach of F. P. Zaostrovsky and K. N. Shabalin developed within the years 1951-1953 and improved further by E. P. Mednikov and other authors [1] -[3] for the case when aerosol deposition was considered as a sub-set of the diffusion process and the equations from the mass transfer theory and the turbulent migration theory were used for determining the aerosol deposition.

EQUATIONS GOVERNING AEROSOL TRANSPORT

Packed gas separators are used for removal of liquid aerosols from gases in cases when there is no solid phase present in the gas. Below is given a mathematical model for the process of removal of liquid aerosol particles from the gas in the apparatus (channel) free from packing and in the apparatus filled with fine random packing.

For evaluating separation efficiency of the packed layer, we will use the equation for convective mass transfer of particles.

Assume that a gas under turbulent conditions containing a finely dispersed phase of concentration C_i is fed to the inlet of the packed layer. As the gas flows, the dispersed phase migrates toward the walls of the packing elements by means of various mechanisms.

The following assumptions are adopted for the mathematical model [1] :

- The wall material of the packing elements is well wetted by the deposited liquid phase.
- The particle diameter d_p is small compared to the scale L of pulsation vortices carrying the particles: $d_p = L$.
- The particle polydispersity is accounted for by considering particles by fractions.
- At $C < 0.2$ kg/m^3, the particles do not collide and do not coalesce with each other.

Then, a three-dimensional equation governing steady-state transport of aerosol particles in the gas takes the form

$$u_x \frac{\partial C}{\partial x} + u_y \frac{\partial C}{\partial y} + u_z \frac{\partial C}{\partial z} = \frac{\partial}{\partial x}\left(D_d \frac{\partial C}{\partial x}\right) + \frac{\partial}{\partial y}\left(D_d \frac{\partial C}{\partial y}\right) + \frac{\partial}{\partial z}\left(D_d \frac{\partial C}{\partial z}\right)$$

(3)

Where x, y, z are spatial coordinates (m); D_d is the turbulent diffusion coefficient for particles (m²/s).

The Brownian diffusion is neglected here, since it plays only a very small role in turbulent flows.

For solving Equation (3), a 3-D velocity field (u_x, u_y, u_z) and a spatial distribution of the coefficient of turbulent transfer of particles $D_d(x, y, z)$ are required. Then, Equation (1) can be solved using the preset concentrations at the boundaries to obtain a spatial distribution of particles along the coordinate axes 0x, 0y, and 0z. As it was noted in [1] , it is not easy to set values of aerosol concentrations at the walls of packing elements due to some uncertainties. Besides, it must be kept in mind that packing elements are distributed randomly within the gas separator volume. Therefore, below we will use an approach to the simulation that considers transfer of the dispersed phase mass toward the wall (or toward the phase interface) as a volumetric source of mass in the particle transport equation.

The source of mass is presented in the general form as follows

$$r = \frac{M}{V} = \frac{jF}{V}$$

(4)

where M is particle flux (kg/s); F is area of contact of gas with the walls of the channel filled with packing (m²); V is working volume of the contact unit (m³); j is the average value of specific particle flux toward the wall at the wall surface (kg/(m²s)).

In the theory of gas treatment, the specific flux of the dispersed phase mass can be written in the form of an analog of the mass transfer equation as follows: $j = u_t C_\infty$, where C_∞ is the particle concentration in the flow core taken as cross sectional area-average particle concentration (kg/m³); u_t is the turbulent migration velocity (m/s). The so-called turbulent particle deposition velocity, u_t, characterizes the intensity of particle deposition from the turbulent flow onto the walls, i.e. it represents the quantity of particles (weight or number) being deposited from the stream containing aerosols onto the walls of area 1 m² per 1 s divided by a unit particle concentration [1] .

With the assumptions made, Equation (3) for a cylindrical channel filled with packing takes the form:

$$u_z \frac{\partial C}{\partial z} = \frac{1}{r} \frac{\partial}{\partial r}\left(rD_d \frac{\partial C}{\partial r}\right) + \frac{u_t C_\infty F}{V}$$

$$(5)$$

Where r is radial distance from the channel's axis (m); z is vertical coordinate (in the direction of the gas flow) (m).

TURBULENT PARTICLE DEPOSITION VELOCITY

A number of empirical and semi-empirical dependencies can be used for calculation of reduced rate of particle deposition $u_t^+ = u_t/u_*$ which is related to particle relaxation time

$$\tau_p = \frac{d_p^2 \rho_p}{18\mu_g} = \frac{d_p^2 \rho_p}{18\rho_g v_g}$$

$$(6)$$

Where d_p is particle diameter (m); ρ_p is density of particle material (kg/m³); v_g is kinematic viscosity of the gas (m²/s); μ_g is dynamic viscosity of the gas (Pa·s); ρ_g is density of the gas (kg/m³).

Dimensionless relaxation time is defined as:

$$\tau^+ = \frac{\tau_p u_*^2}{v_g}$$

$$(7)$$

where u_* is friction velocity on the wall (m/s).

Generalized expression for u_t takes the form [1] :

$$u_t = 7.25 \times 10^{-4} \left(\frac{\tau^+}{1 + \omega_E \tau_p} \right)^2 ,$$

(8)

Where $\omega_E = u_*/(0.1R)$ is angular velocity of energy-intensive pulsations (1/s); R is the channel radius (m).

From Equations (7) and (8) it follows that separation rate u_t depends to a large extent on friction velocity u_* or wall shear stress τ_w ($u_* = (\tau_w/\rho_g)^{0.5}$).

The average value of friction velocity on the surfaces of random packing elements can be determined through using the average energy dissipation rate expressed through pressure drop.

Numerous calculations of u_* carried out for packed separators agree with accuracy ±18% with the expression [3] :

$$u_* = 1.8 \left(\frac{\bar{\varepsilon} v_g}{\rho_g} \right)^{\frac{1}{4}} = 1.8 \left(\frac{\Delta p u_{ave}}{l \rho_g \varepsilon_{free}} \right)^{\frac{1}{4}} ,$$

(9)

Where the average energy dissipation rate is determined through the pressure drop as follows:

$$\bar{\varepsilon} = \frac{\Delta p S u_{ave}}{V_{free}} = \frac{\Delta p u_{ave}}{l \varepsilon_{free}} ,$$

(10)

where V_{free} is free volume of the packed layer (m³); S is cross-sectional area of the packed layer (apparatus) (m²); l is length (height) of the packed layer (m); ε_{free} is specific free volume of the packing, where u_{ave} is average velocity of gas flow in the packing (m/s); $u_{ave} = w_g/\varepsilon_{free}$; w_g is gas flow velocity in the channel free from packing (m/s).

TURBULENT DIFFUSION COEFFI-CIENT FOR PARTICLES

Turbulent diffusion coefficient for particles can be determined using the following expression [1] :

$$D_d = \frac{v_T}{1 + \omega_E \tau_p},$$

(11)

Where v_T is the turbulent viscosity coefficient in a single-phase flow (m²/s);

For a channel free from packing, the average value of v_T at $y > \delta$ can be calculated using the Karman–Prandtl model:

$$v_T = \chi u_* \delta,$$

(12)

Where $\chi = 0.4$ is a turbulence constant; δ is the average thickness of the near-wall layer (m).

Turbulent viscosity coefficient in the packed layer is determined using the following semi-empirical expression [4] :

$$v_T = 3.87 v_g \sqrt{Re_e \, \xi},$$

(13)

Where Re_e is the Reynolds number defined as $Re_e = u_{ave} d_e / v_g$; is the hydraulic resistance coefficient for the packing [3] .

NUMERICAL RESULTS

Results of the numerical solution of Equation (5) by the finite-difference method are shown below.

Figure 1 presents concentration profiles along the channel length I for a hollow tube and for a tube filled with steel Raschig rings. The gas-liquid mixture is air containing water droplets. Initial data are following: d = 0.017 m; C_i = 5 wt%; u_{ave} = 30 m/s; d_p = 9 μm. Apparently, a tube filed with packing ensures better purification efficiency compared to a hollow tube.

Figure 2 shows change in separation efficiency η with the channel length l for different Reynolds numbers. In the calculations, a tube filled with random packing was assumed. The contact devices are the Raschig rings. The gas-liquid mixture is air containing water droplets. Initial data are following: d = 0.017 m; C_i = 5 wt%; d_p = 9 μm. It can be seen that the smaller the channel length is, the greater the Reynolds number must be for high purification efficiency. Conversely, as the channel length increases, the Reynolds number can be reduced without loss in separation efficiency. Figure 3 shows change in separation efficiency η with the Reynolds number for different channel lengths l. The tubes are assumed to be filled with steel Raschig rings. The gas-liquid mixture is air containing water droplets. Initial data are following: d = 0.017 m; C_i = 5 wt%; d_p = 9 μm. It can be seen from the figure that the increase in the channel length and the Reynolds number leads to an increase in separation efficiency. Figure 4 shows change in separation efficiency η with the Reynolds number for particles of different sizes.

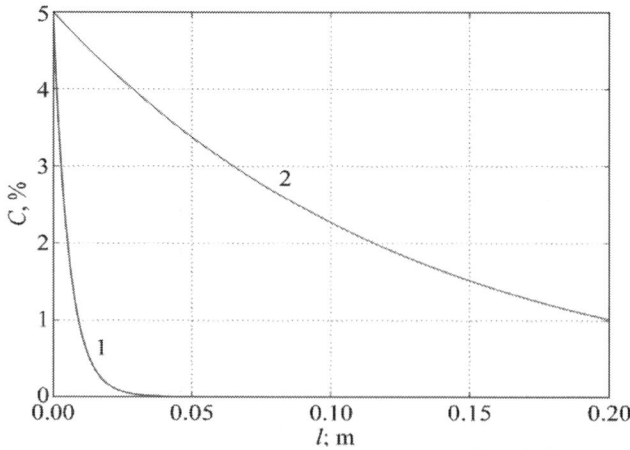

Figure 1: Concentration profiles for a tube filled with Raschig rings and a hollow tube: 1—Tube filled with packing; 2—Hollow tube.

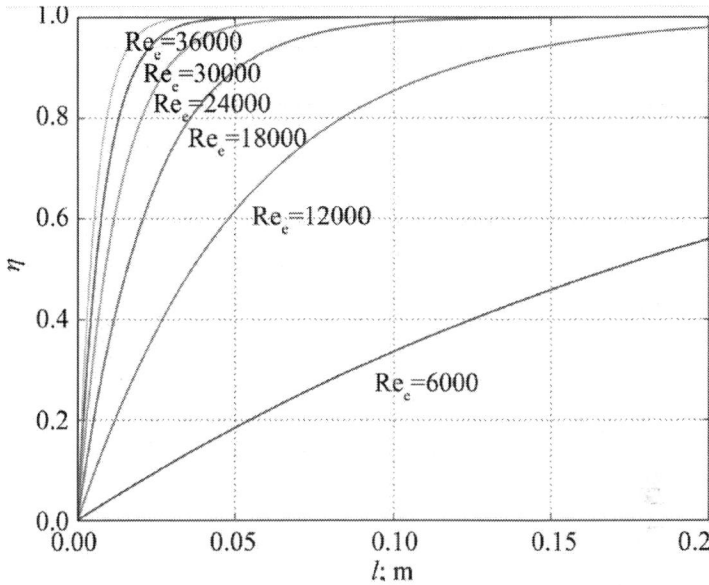

Figure 2: Change in separation efficiency with the channel length for different Reynolds numbers.

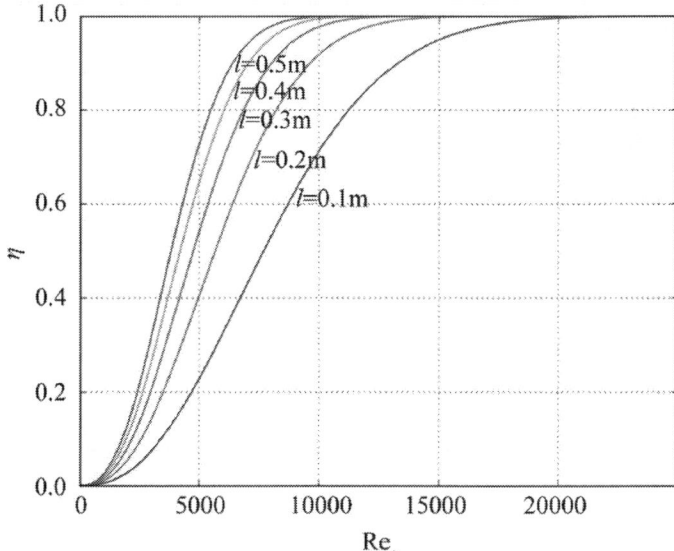

Figure 3: Change in separation efficiency with the Reynolds number for different channel lengths.

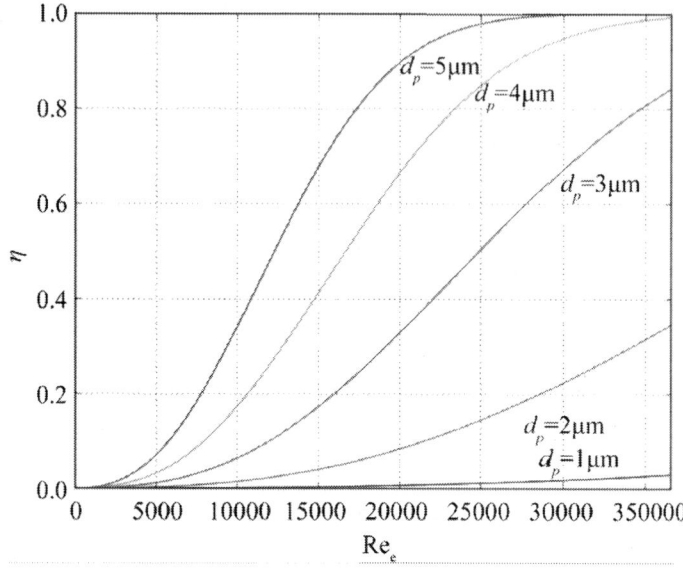

Figure 4: Change in separation efficiency with the Reynolds number for particles of different sizes.

The tubes are assumed to be filled with steel Raschig rings. The gas-liquid mixture is air containing water droplets. Initial data are following: d = 0.017 m; C_i = 5 wt%; l = 0.2 m.

Apparently, the increase in the particle size and the Reynolds number leads to an increase in separation efficiency.

If the Peclet number $Pe_e = u_{ave}l/D_{mix} > 10$, the model of ideal displacement in the channel can be used, where D_{mix} is mixing coefficient (m²/s).

The mixing coefficient can be calculated using the following empirical expression [4] .

$$D_{mix} = 1.92\nu_g \, Re_e^{0.75} \, \xi^{0.25} .$$

(14)

As is known, for $Pe_e > 10$, solution to Equation (5) takes the form

$$\eta = \frac{C_i - C_f}{C_i} = 1 - \exp\left(\frac{4lu_t}{u_{ave}d_e}\right) .$$

(15)

Calculations carried out with the use of the above expression agree satisfactorily with experimental data [1] -[6] .

CONCLUSIONS

The considered approach to determination of particle concentration profiles in contact devices and to calculation of separation efficiency which is based on the turbulent migration theory allows carrying out calculations with the use of hydraulic resistance of the apparatus. The obtained equations can be used for calculating industrial gas separators present at petrochemical and gas producing enterprises.

REFERENCES

1.   Mednikov, E.P. (1980) Turbulent Transport and Deposition of Aerosols. Nauka, Moscow.

2.   Sugak, E.V., Voinov, N.A. and Nikolaev, N.A. (2010) Cleaning of Off-Gases in Apparatuses with Intense Hydrodynamic Regimes. ZAO Novoe Znanie Press, Kazan.

3.   Laptev, A.G. and Farakhov, M.I. (2006) Separation of Heterogeneous Mixtures in Packed Apparatuses. Kazan State Power Engineering University Press, Kazan.

4.   Laptev, A.G., Farakhov, T.M. and Dudarovskaya, O.G. (2013) Models of Turbulent Viscosity and Mixing in Channels and Packed Flow-Through Mixers. Russian Journal of Applied Chemistry, 86, 1046-1055. http://dx.doi.org/10.1134/S1070427213070173

5.   Laptev, A.G. and Basharov, M.M. (2013) Efficiency of Aerosol Deposition in Gas Separators of Various Designs. Russian Journal of Applied Chemistry, 86, 1190-1196.http://dx.doi.org/10.1134/S1070427213080077

6.   Laptev, A.G. and Basharov, M.M. (2013) Determination of Efficiency of Removal of Disperse Phase from Gases by Vortical Elements. Izvestiya VUZov Khimiya i Khimicheskaya Tekhnologiya, 56, 101-104.

Chapter 2

Safety of Lebanese Bottled Waters: Vocs Analysis and Migration Studies

Aline Ghanem, Jacqueline Maalouly, Roula Abi Saad, Dalia Salameh, and Chawki Oscar Saliba

UMR (Emballages Alimentaires au Liban), Faculty of Sciences II, Lebanese University, Jdeidet El-Matn, Lebanon

ABSTRACT

Despite the worldwide increase in the consumption of PET-bottled mineral waters compared to tap waters encouraged by its microbiological and chemical safety for public health, contaminants could migrate from the plastic packaging into the water and induce adverse effects on human health. Volatile organic compounds (VOCs), including benzene, toluene, ethylbenzene, ortho, Meta, and para-xylenes (BTEX), styrene, chlorobenzene and benzaldehyde are among the potential contaminants of bottled waters. This study aimed to assess Lebanese PET-bottled waters, in respect of VOCs contents, with comparison to polycarbonate-bottled and tap waters. Both HS-SPME-GC/FID and SPE-GC/FID were optimized and validated for VOCs determination in the waters, and their performances were compared. The HS-SPME-GC/FID was valid (Afnor NF T 90-210 (May 2009)) for all the studied molecules with limits of quantifications (LOQ) far lower the maximum contaminants levels (MCLs) set by both US-EPA and WHO. SPE-GC/FID was valid only for ethylbenzene, m/p-xylenes, o-xylene, and styrene, with poorer LOQs. HS-SPME-GC/FID was used therefore for VOCs monitoring in studied water samples showing the safety of the Lebanese bottled-water. The effects of bottles storage conditions (time, and exposure to sunlight) on VOCs migration were also studied.

INTRODUCTION

Bottled drinking water consumption is increasing especially in developing countries, due to safety reasons, because in these countries tap water is rarely submitted to control and monitoring to ensure its chemical and microbiological quality. Various contaminants including aldehydes and phthalates occur in bottled ad tap waters [1-3]. Volatile organic compounds (VOCs), such as benzene, toluene, ethylbenzene, ortho, meta, and para-xylenes called (BTEX), as well as styrene, chlorobenzene and benzaldehyde may also be present in bottled and tap waters [4] due to: 1) their relatively high solubility in water which is 18, 25, 3, 20, 50, 100, 300, 500 and 4600 mg·L^{-1}, respectively [5], 2) their non-intentionally addition to PET during its production and therefore their potential migration from bottles to waters, and to 3) the environmental contamination from industrial activities of springs, ground and surface waters from which drinking waters are derived. The contamination of the aforementioned environmental matrices by VOCs is one of the consequences of the environmental pollution by waste solvents, oils and sewage, gasoline cars and pesticides, which leads to the VOCs transfer to drinking waters [5,6].

Another source of these VOCs in bottled mineral waters is the plastic packaging. The polyethylene terephthalate (PET) is the most commonly used especially for making water bottles below 7 liters. This polymer is obtained by polymerization of ethylene glycol (MEG) with terephthalic acid (TPA) or dimethylterephtalate (DMT). After this first step, a second polycondensation is carried out with antimony-, germaniumor titanium-based catalyst [7,8]. Then the PET obtained as granules is submitted to an extrusion process at 285°C to produce preforms, and subsequently to a blow molding process between 95 and 100°C, which gives the bottle its final form [9]. Polycarbonate (PC) is used to make water containers above 7 liters. Most migration studies concerned the migration of bisphenol A (BPA) from PC bottles [10,11]; however, other compounds might be present in PC-bottled food and waters and deserve therefore an investigation.

BTEX are used worldwide as solvents and chemical intermediates in plastic production [12]; whereas styrene derived from composites of PET with polystyrene (PS) [13]. Residual traces of these VOCs might migrate from PET bottles to the bottled water. Otherwise and during injection-blowing, high-temperature is able to cause the degradation of PET and the appearance of by-products including VOCs which can leach out, by diffusion into the water after bottling [3]. Similarly it is shown that the rate of the secondary products in PET increases exponentially with the applied temperature, especially for benzene [14]. Also the presence of oxygen [15] and humidity can promote degradation reactions [16]. Finally the migration of these pollutants from PET to bottled water is held according to the partition and the diffusion coefficient of each compound between the two mediums. Chemical migration increases with the storage time, the exposure to UV light and to the high temperature.

Dutra et al. [17] have shown the presence of benzaldehyde in PET virgin bottles without being present in the granules before their formatting. This may be due to a degradation of the polymer during the process. In addition the detection of this VOC by several studies in the recycled PET [18,19] allows us to consider that its presence in bottled mineral water is an indicator that the used PET is totally or partly recycled, because benzaldehyde is used as solvent for printing inks used for bottles labelling.

Today, the water contamination with VOCs is a topical issue challenging not only the scientific community for developing suitable and sensitive analytical methods, but is worrying as well for environmental, social and public health concerns. Due to the toxicity (carcinogenic or mutagenic properties) of the VOCs even at trace levels, many countries focus their research on the monitoring of these contaminants in waters [20].

Food contact plastic materials are covered by the European regulation N° 10/2011 which establishes the list of authorized compounds for use in plastic formulation and sets restrictions for individual authorized substances expressed as specific migration limits (SML) [21]. Also the US-Environmental Protection Agency (EPA) [22] and the World Health Organization (WHO) [23] set maximum concentration levels (MCLs) for the potential migrants in drinking water. These values are relatively low for the VOCs, especially for benzene which are of 5 and 10 $\mu g \cdot L^{-1}$ for EPA and WHO, respectively.

The Purge and trap-GC/MS technique is adopted by the U.S.EPA 524.2 method [24] as the most widely used for the analysis of VOCs in water, allowing low detection limits (LOD) of 0.1 $\mu g \cdot L^{-1}$ for BTEX and styrene, far below their MCLs [4,25]. Flame ionization detector (FID) allows less sensitivity with LODs ranging between 0.3 and 2.4 $\mu g \cdot L^{-1}$ [7]. The use of photoionization detector (PID) shows LODs between 5.6 and 7.2 $\mu g \cdot L^{-1}$ [26]. However the drawbacks of the Purge and trap are the requirement of complex instrumentation, the interference of water vapor generated at the purge stage, the possible contamination of the trap (cross contamination), longer analysis time per sample, and its nonavailability in all analytical laboratories.

HS-GC/MS arises as an alternative method with a LOD of 0.02 $\mu g \cdot L^{-1}$ [27]. The replacement of the MS by FID or PID detector allows LOD between 0.2 and 7.5 µg·L^{-1} which is not adequate for all VOCs [28]. Head space solid phase dynamic extraction (HS-SPDE) coupled to GC/MS is a very efficient technique with a LOD of 20 ng·L^{-1}, but it is not available in all laboratories [29]. Another fast method is the direct aqueous injection DAIGC/MS that allows LOD ranging from 0.6 to 1.1 µg·L^{-1} [30,31]. This technique requires the use of an absorbing material in the liner of the injection port and a temperature below 70°C to the injector, to prevent the passage of water to the separation column. But this temperature is not adequate for the separation of VOCs with a higher boiling point.

Solid phase extraction (SPE) is a widely used technique for sample preparation including purification and concentration in a single step with the use of a small volume of solvent eluent. SPE-GC/MS gives a LOD between 6 and 10 µg·L^{-1} [32], because of the loss of VOCs by volatility during sample concentration [33]. In the literature, some authors suggest the conservation of the sample and the elution solvent at 4°C, or the addition of a Cetrimonium bromide (CTAB) surfactant in the sample, to limit the loss of VOCs by formation of micelles and subsequently ensure a good recovery [34].

The solid phase micro extraction (SPME) is used as an alternative extraction technique offering the advantages to be simple, fast and does not require solvents (green and environmentally friendly technique). In immersion and head space mode, the LOD is between 0.01 and 0.5 µg·L^{-1} with GC/MS [17,35-37] and between 0.01 and 1 µg·L^{-1} with GC/FID [38-41]. However the HS mode has proven to be more advantageous because it often shows an important reduction of extraction time [41].

The objective of our work is to verify the safety of mineral bottled waters in Lebanon in respect to the VOCs contamination. Analytical methods involving HS-SPMEGC/FID and SPE-GC/FID were developed and their performances were compared. Optimization of the parameters of each technique was accomplished by using the methodology of the experimental design [42]. Methods validation was carried out following the Afnor NFT 90 - 210 (May 2009) criteria. The methods were applied to monitor VOCs in Lebanese bottled water samples with comparison to tap waters in order to assess the safety of the drinking water in our country. A study of the VOCs migration from the PET to the bottled waters was carried out involving the storage time, and the exposure to sunlight.

MATERIALS AND METHODS

Chemicals and Standards

Ethylbenzene (98%), m/p-xylene (99%), styrene (99%), o-xylene (98%), fluorobenzene (99%) (Internal Standard, IS) and benzaldehyde (98%) were purchased from Sigma-Aldrich (Steinheim, Germany); benzene (extra pur) was purchased from Merck (Germany); toluene (99.7%) from Fluka Analytical (St. Louis, MO, USA) and chlorobenzene (99%) from Riedel de Haën (Seelze, Germany). All chemicals and solvents used were of analytical-reagent and chromatographic grade, respectively. The 0.45 µm filtered ultrapure water (conductivity 0.055 µS/cm) was used for solutions preparation and was generated by "TKA" Smart 2 Pure Water System (Niederelbert, Germany). Stock solutions of mixture compounds were prepared in acetonitrile at 500 mg·L^{-1}. The working standard solutions of all volatile organic compounds were prepared by diluting the stock solution in water.

Sample Collection

Different samples of mineral bottled water were collected from the Lebanese market. Sampling was performed mainly according to the following criteria: bottle brand and type (PET, PC, or glass), bottle volume (0.33, 0.5, 1, 1.5, 2, and 6 liters for PET, 18.9 liters for PC and 0.75 liters for glass). Comparison was made with tap water. All samples were analyzed within their validity date (before expiry date as mentioned on the bottle), during the year 2010-2011.

In addition, the effect of storage mode of PET-bottles on the migration of VOCs into the bottled water was also evaluated. For this part of study, PET-bottled water samples from different brands and sizes were collected in duplicate and were divided into two equivalent groups (same batch number) to be stored during 5 months before analysis: one was stored at dark at 20°C as mean temperature and one in open air under sunlight (35°C as mean temperature).

To assess the quality of the PET-bottled water samples at their expiry date (Directive 2002/72/EC), a series of PET-bottled samples (0.5 L) were analyzed freshly after collection and 10 similar samples were incubated at 40°C during 10 or 20 days, then analyzed and the results were compared.

Analytical Procedures

For HS-SPME, a 10 mL of sample solution containing 10 $\mu g \cdot L^{-1}$ of fluorobenzene as internal standard was undertaken, added to a 20 mL sample vial equipped with a PTFE-coated septum. A 65 μm PDMS/DVB fiber was used to extract the VOCs from water samples. Duplicate analyses were performed under the following conditions: extraction temperature (25°C), extraction time (31 min), NaCl concentration (0.2 $g \cdot mL^{-1}$), and sample solution stirring.

SPE extraction was carried out on a Visiprep™ SPE Vacuum Manifold (Sigma-Aldrich, Steinheim, Germany). The SampliQ C18 SPE cartridges (500 mg, 6 mL) provided from Agilent (USA) were conditioned by flushing 10 mL of methanol followed by 10 mL of ultrapure water. Water samples (330 mL) were loaded on the cartridges at a flow rate of 3 $mL \cdot min^{-1}$. The cartridges were then rinsed with 3 mL of water and analytes were eluted in 10 mL of acetonitrile containing 108 μL of dodecane, added as keeper to decrease the rate of VOCs volatilization. Sample concentration was then carried out under a stream of nitrogen until 50% of evaporation of the extract volume. After SPE concentration, 1 μL of the sample were injected in the GC injector set on splitless mode. Chromatographic separation was performed on a gas chromatograph

(Agilent technologies 6890N Series N.05. 04) equipped with FID detector. The data acquisition and processing were done using Agilent Chem Station software. The separation of BTEX (benzene, toluene, ethylbenzene, and xylenes), styrene, chlorobenzene and benzaldehyde was carried out on a DB-5 column (30 m × 0.25 mm i.d.) with 1 μm of film thickness. The GC oven temperature was programmed as follows: 50°C held for 3 min, rate at 15°C/min to 90°C and held for 3 min; rate 15°C/min to 60°C and held for 2 min and finally rate 15°C/min to 140°C and held for 4 min. The total time for each GC run was 22 min. Helium, at a constant flow rate of 2 mL/min, was used as the carrier gas. The injector and detector temperatures were set at 250°C and 300°C, respectively. For HS-SPME, the injection port fitted with a 0.75 mm i.d. injection liner (Supelco, Sigma-Aldrich, Steinheim, Germany) was operated in the splitless mode.

Methods Validation

The linearity and limits of quantification were validated according to standard AFNOR NF T 90 - 210 (May 2009) [43], under the optimum conditions of both HS-SPME and SPE-GC/FID methods. This validation was obtained by extracting five different concentrations of standards ranging between LOQ and 100 $\mu g \cdot L^{-1}$, with fluorobenzene (10 $\mu g \cdot L^{-1}$) as internal standard for HS-SPME; whereas standards concentrations were 50, 70, 90, 120 and 250 $\mu g \cdot L^{-1}$, with chlorobenzene (90 $\mu g \cdot L^{-1}$) as internal standard, for SPE. Calibrations were performed on five consecutive days.

According to the AFNOR document NF T 90 - 210 (May 2009), 10 measurements of independent solutions with a concentration at the estimated limit of quantification LOQ were performed (mean = zLOQ, standard deviation sLOQ). The estimated limit of quantification is acceptable when zLOQ − 2 × sLOQ > LOQ − 60% × LOQ and zLOQ + 2 × sLOQ < LOQ + 60% × LOQ. The 0.6 value has been defined by convention but can be modify in agreement with specific regulations. To ensure the quality control of the analysis, the samples analysis were performed after verifying the absence of memory effect in the chromatograms. All sample analysis were performed in duplicate, and a blank sample was analyzed every six sequences.

Experimental Designs

HS-SPME Design

To Central Composite Designs (CCDs) were used to combine the experiments of a factorial design (Nf), the star points (2k + 1) and N0 in the center of the field trials. This is a quadratic model with $N = 2 \wedge k\, 2 + k\, N0$, k is the number of factor. The star points are located at $\pm\alpha$, with $\alpha = 1$ in this case the center of the experimental domain. These tests are on the faces of the cube [44]. The number of experiments was 30 involving 4 factors (Table 1) and 6 replicates of the central point were included to assess the experimental errors.

This design allows assessing isoresponse curves. Once the experiments and the model coefficients are calculated, we obtain the response surfaces shown in Figure 1. To find an acceptable compromise area for the experimental conditions we have defined the individual desirability functions for the studied contaminants. These are left unilateral functions, knowing that in the global desirability, no preponderance was imposed.

All "blank" water samples were spiked with the standards in mixture at 10 $\mu g \cdot L^{-1}$ for each compound. CCD was used to optimize the four factors (cited with their low (−1), central (0), and high levels (+1)). The optimized factors were: 1) the ionic strength by means of the NaCl concentration (0 - 0.1 - 0.2 $g \cdot mL^{-1}$), 2) extraction temperature (25°C - 50°C - 75°C), 3) extraction time (5 - 18 - 31 min), and 4) headspace volume (ratio phase of sample volume in 20 mL vial) (2/20 - 6/20 - 10/20).

Using the CCD, the peak areas of all analytes were extracted from the chromatograms and were considered as the responses. The software used is NemrodW (LPRAI, Marseille-France, 2011), which allows us to draw the appropriate matrices, to calculate the coefficients of the model, to represent the response surface, the desirability surface and to find the optimum.

SPE Design

Similar CCD was used for SPE optimization (Table 2) and involved 30 experiments and 4 factors that are: 1) sample volume (250 - 350 - 450 mL), 2) elution volume of acetonitrile (10 - 14 - 18 mL), 3) keeper volume in elution phase (62 - 93 - 124 μL) and 4) percentage of the evaporation of the extract volume under nitrogen stream (50% - 60% - 70%). The water samples were spiked with standards in a mixture at 250 $\mu g \cdot L^{-1}$ for each compound. The extraction recoveries (calculated from the peak areas obtained) were considered as the responses to include in the chemometric calculations (Figure 2).

RESULTS AND DISCUSSION

Optimization of the HS-SPME

Choice of the Fiber Coating

In order to select the best fiber to extract VOCs, several fiber coatings purchased from Supelco (Sigma-Aldrich, Steinheim, Germany), were tested: 1) a fused silica fiber coated with 100 μm layer of polydimethylsiloxane (PDMS), 2) a fused silica fiber partly coated (composite coatings) with a 65-μm layer of polydimethylsiloxane/divinylbenzene (PDMS/DVB) [36], and 3) a fused silica fiber partly coated with a 75-μm layer of carboxen/polydimethylsiloxane (CAR/PDMS) [45].

A test of repeatability (n = 3) was realized to choose the most repeatable fiber, using water samples spiked with the mixture of standards at 10 $\mu g \cdot L^{-1}$ for each compound. In a vial of 20 mL,

10 mL of spiked water with a NaCl concentration of 0.2 g·mL^{-1} (HeadSpace HS ratio of 10/20) were extracted under magnetic stirring for 10 min at 25°C. The fiber was then introduced into the GC injector to desorb at 250°C during 1.5 min the analytes from the fiber coating. The relative standard deviation (RSD%) of the absolute area of each analyte was determined.

The results obtained (Table 3) showed that the PDMS/ DVB coating allows the best peak area repeatability of analytes with RSD lower than 16%; whereas the RSD values reaches 35% with PDMS for the most volatiles (i.e. benzene and toluene) and 77% with CAR/PDMS for the benzene. The non-polar PDMS coating is recommended by the supplier for non-polar with low molecular weight compounds such as alkanes, etc., via adsorption [45]. However, our results showed poor repeatable extraction with the PDMS when dealing with high volatiles compounds, which could be explained by unreached partition equilibrium of the analytes between the polymeric stationary phase and the sample matrix.

Despite that CAR/PDMS, a bipolar adsorbent, has been previously described for BTEX analysis at trace levels by SPME coupled to cryo-trap GC/MS [45], this fiber coating was poorer repeatable for benzene (the most critical analyte) extraction when HS-SPME-GC/FID is involved. Our analytes were adsorbed more efficiently, and released faster with more repeatable extraction, on the 65- μm (PDMS/DVB)-coated fiber. With the thinner coating and higher specific surface area, the PDMS/DVB extracts more of our analytes through diffusion and better release the compounds during thermal desorption.

Table 1: Experimental variables, levels, matrix of CCD and peak areas (a.u.) for the VOCs by HS-SPME-GC/FID using the PDMS/DVB fiber

| | Extraction conditions | | | | Peak areas | | | | | | | | |
Run	NaCl (g/ml)	Time (min)	Temperature (°C)	Volume HS (mL)	Benzene	I.S.	Toluene	Chlorobenzene	Ethylbenzene	m/p-Xylenes	Styrene	o-Xylene	Benzaldehyde
1	-1	-1	-1	-1	7.6	5.1	7.4	7.7	9.4	24.2	15.2	12.6	1
2	+1	-1	-1	-1	6.6	11.5	15.4	14.9	15.3	40.8	23.8	22.5	2.9
3	-1	+1	-1	-1	3.4	5.6	15.7	17.9	23	51.1	28.4	26.1	4.4
4	+1	+1	-1	-1	6.3	16.4	20.7	32.7	29.2	73.2	33.5	35	11.4
5	-1	-1	+1	-1	0.31	3.4	1.3	2.1	3.3	8.2	6.9	5.9	3.4
6	+1	-1	+1	-1	0.24	11.9	1.2	3	5.3	10.6	8.5	7	5.5
7	-1	+1	+1	-1	0.13	7.4	1.1	2	3.2	8.2	7.4	6.4	3.3
8	+1	+1	+1	-1	0.23	5.8	1.6	3.6	5	13.1	10.6	9.4	10.3
9	-1	-1	-1	+1	5.7	9.2	16.8	16.5	24.9	63.5	22.1	25.3	1
10	+1	-1	-1	+1	9.9	13.8	25.7	26.2	31.9	82.6	37.9	39.4	1.7
11	-1	+1	-1	+1	6.7	7.8	31.9	43	50.3	149	80	110.9	4.4
12	+1	+1	-1	+1	10.1	19.4	41.5	52.8	50.3	128	82	72.4	9.5
13	-1	-1	+1	+1	0.55	6.6	3.7	10.5	17.3	42.5	29.8	26.7	3.3
14	+1	-1	+1	+1	1.4	4.6	10.5	16.8	24.7	69	45.7	39.2	10.7
15	-1	+1	+1	+1	0.86	1.4	8.9	12.3	22.4	58.3	37.2	34.6	4.8

16	+1	+	+1	+1	1.4	3.3	10.6	17.1	25.4	75.6	49.1	44.2	12.5
17	-1	0	0	0	2	5.1	15.4	29.7	38.9	100.5	62.3	56.3	7.1
18	+1	0	0	0	6.4	5.9	18	32.2	41.7	130.8	74.6	57.6	16.4
19	0	-1	0	0	2.8	5.5	15.6	25.3	33.8	87.1	48.1	45.5	4.6
20	0	+1	0	0	8.3	4.5	9.5	19.6	26.3	79	59.9	43	11.8
21	0	0	-1	0	10.1	17.7	38.8	41.2	42.5	114.2	62.5	60.6	3.6
22	0	0	+1	0	2	12.1	4.5	8.4	13.3	42.6	34.3	23.5	6.4
23	0	0	0	-1	1.1	4.2	7.2	10.6	16.1	41.2	23.4	22.2	16.1
24	0	0	0	+1	6.6	12.7	28.3	45	59.6	161.2	84.7	83.4	13.8
25	0	0	0	0	2.7	9.1	15.9	32	46.9	121.8	71.6	65.2	10.8
26	0	0	0	0	3.9	3.8	20	34	40.3	116.7	72.2	59	14.5
27	0	0	0	0	6	4.1	15.4	30.2	39.6	108.1	69.9	54.7	14.8
28	0	0	0	0	2.3	4.8	15.7	25	41.8	110	67.1	60.2	14.1
29	0	0	0	0	4.6	4.4	17.1	33.5	38.8	110.6	70.5	56.3	14
30	0	0	0	0	3.8	8.6	24.6	40.4	51.9	127	71.4	66.4	14.7

Optimization Results

The optimization of the HS-SPME was performed on the fiber that allows the best repeatability (65 μm PDMS/ DVB). The desorption temperature and time of the VOCs from the SPME fiber into the GC injector were optimized classically by testing desorption efficiency of the analytes by means of their peak areas in the corresponding chromatograms, with the absence of memory effect in the further blank injections.

Table 2: Experimental variables, levels, matrix of CCD and yields (%) for the VOCs by SPE-GC/FID

| SPE variables | | | | | | Yields % | | | | |
Run	Sample volume (mL)	Elution volume (mL)	Keeper volume (µL)	Evaporation under N2 (%)	Fluorobenzene	Toluene	EthylBenzene	m/p-Xylenes	Styrene	o-Xylene
1	250	10	62	50	18.16	88.6	45.2	49	64.32	58.6
2	450	10	62	50	7.2	54.65	37.7	41.2	56.5	48
3	250	18	62	50	12.38	101.3	44.8	48	69.6	51.6
4	450	18	62	50	18	64.4	35.8	37.7	59.2	45
5	250	10	124	50	5.52	112.5	37.9	37	58.3	45.1
6	450	10	124	50	6.84	60	35.75	38	56	39.4
7	250	18	124	50	11	145	30	28	50	38.7
8	450	18	124	50	6.18	52.5	37.8	42.3	61.1	46.6
9	250	10	62	70	8.2	34.08	32	37	52	36.8
10	450	10	62	70	1.87	29.4	30.6	34.8	56	43.5
11	250	18	62	70	0	32.4	32	35.3	56.1	40.7
12	450	18	62	70	0	45	24.05	29.5	45	38.75
13	250	10	124	70	5.3	65.8	24	26.8	46.7	35.2
14	450	10	124	70	0	25	21	23.8	44.5	33.2
15	250	18	124	70	11.7	54.7	23	21.6	45.7	31
16	450	18	124	70	2.47	54	29.2	32.4	54.7	40.3

17	250	14	93	60	0	58.3	30	31	55	37.3
18	450	14	93	60	6	38	23.7	25.7	48	30.2
19	350	10	93	60	7.15	40.5	35	38.2	62.7	44.4
20	350	18	93	60	5.67	97	35.8	38.75	62.6	45.3
21	350	14	62	60	2.47	45.6	30.8	32.7	50.4	37.5
22	350	14	124	60	4.45	60	34	36.6	53	39
23	350	14	93	50	6.9	56.3	39.6	44.3	62.5	49.5
24	350	14	93	70	0	29	25.5	30	46.5	37.5
25	350	14	93	60	4.15	42.5	35	40	57	51
26	350	14	93	60	5.4	62	32.5	38.6	57	46
27	350	14	93	60	10.7	77.5	33.2	38.4	52	38
28	350	14	93	60	2.17	45	31	38	55.2	43
29	350	14	93	60	1.4	63.4	31.2	35	49	34.3
30	350	14	93	60	6.5	45.8	34	39.5	60	47

The studied four HS-SPME factors (salinity, extraction time and temperature, and HS volume) are often described in the literature [36,41,45] as the most relevant. Before optimization, a screening design similar to this described in the literature [36] was applied to assess the pertinence of the aforementioned factors, involving additional factors such as the SPME mode (HS vs direct immersion), the sample pH, and the sample stirring during extraction. From results obtained (data not shown), the HS mode, magnetic stirring with no pH-adjustment of the sample were fixed throughout the experiments.

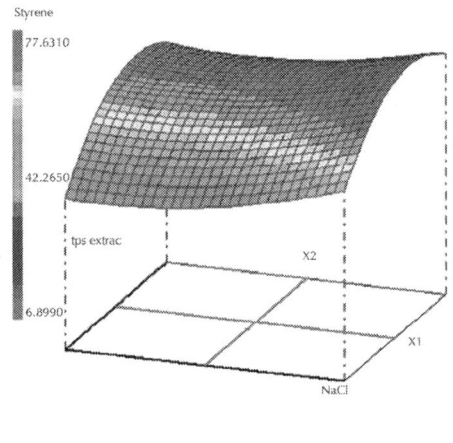

(a)

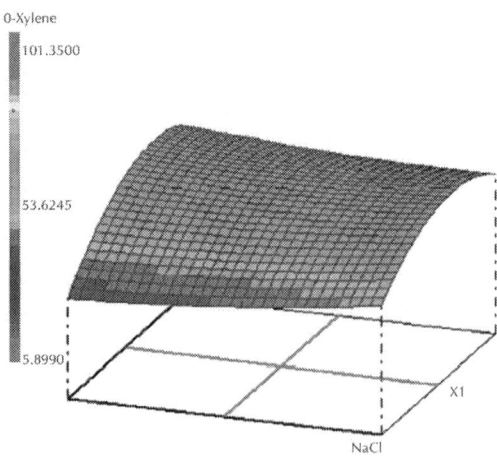

(b)

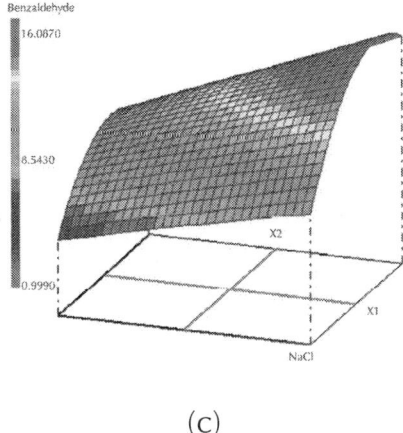

(c)

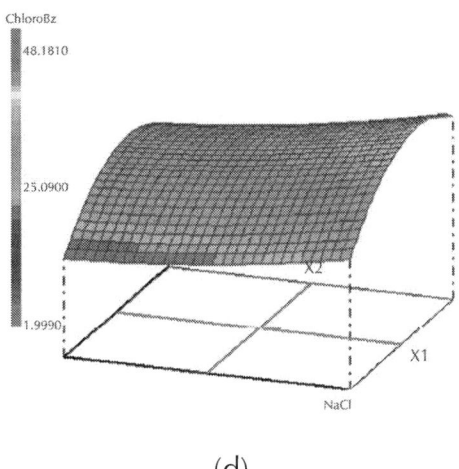

(d)

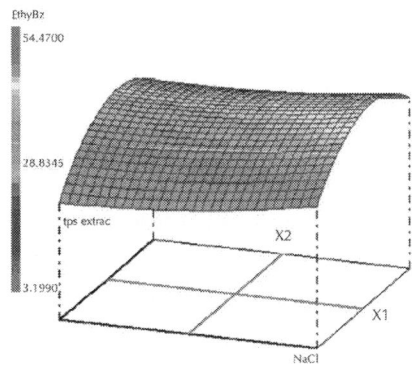

(e)

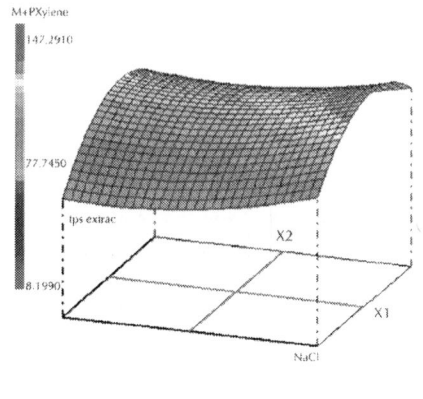

(f)

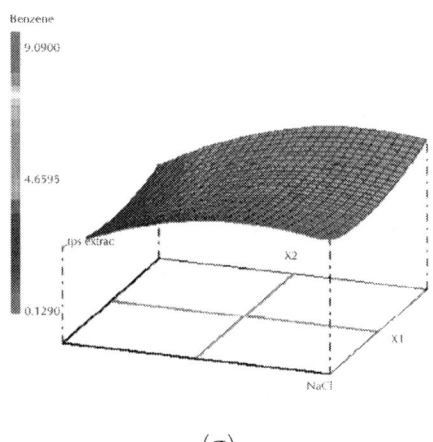

(g)

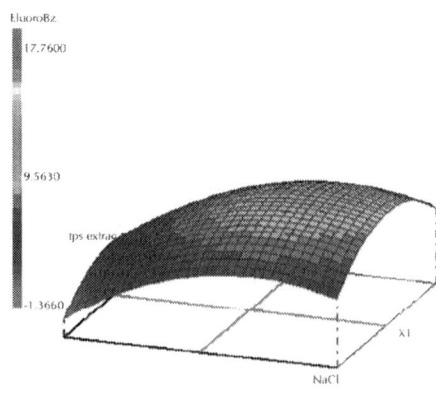

(h)

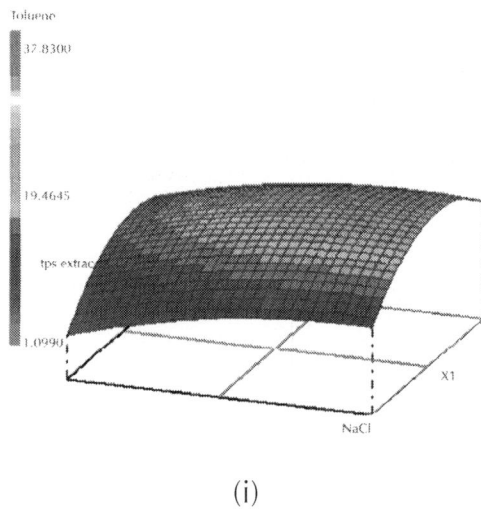

Toluene

37.8300

19.4645

tps extrac

1.0990

X1

NaCl

(i)

Figure 1: Response surfaces obtained by HS-SPME-GC/FID of (a) benz-aldehyde; (b) o-xylene; (c) styrene; (d) m, p-xylene; (e) ethylbenzene; (f) chlorobenzene; (g) toluene; (h) fluorobenzene; (i) benzene.

Optimum HS-SPME conditions were found to be: 0.2 g·mL^{-1} of NaCl; extraction during 31 min at 25°C; Headspace volume of 10 mL in a 20 mL vial. The salting-out effect was obvious in increasing the analytes extraction during an acceptable time. Heating the sample was supposed to increase the VOCs volatility and therefore their transfer to the fiber from the vapor phase. However, the results show that high temperature provokes early desorption of the analytes from the fiber coating during HSSPME extraction, mainly for the most volatile compound (such as benzene). A temperature of 25°C was found to be optimum to extract the analytes with higher efficiency when sample volume is 10 mL in a 20 mL-vial. When the sample volume increases, the mass of analytes transferred to the vapor phase increases leading to higher adsorbed amounts on the fiber coating and therefore to a better extraction.

Global desirability of SPME (DSPME = 0.8004) value was satisfactory [42,46,47]. This function takes into consideration all of the individual desirabilities of all studied contaminants. It is a multiplication of these late functions, if any of these is none satisfactory (= 0), the

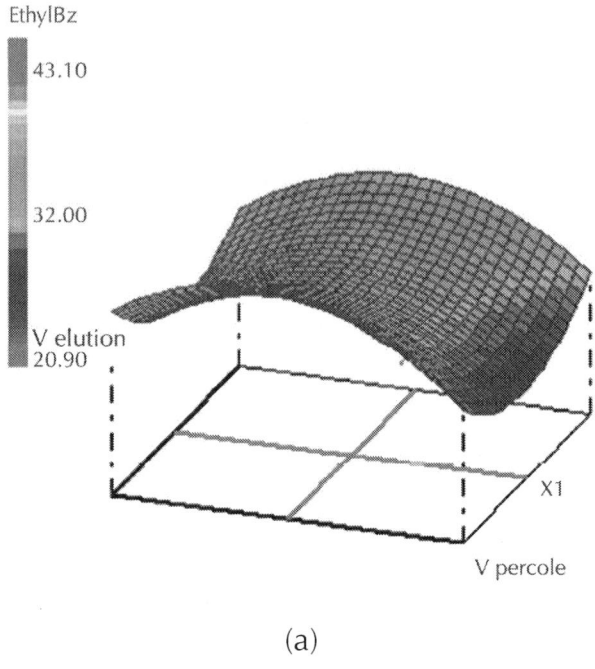

(a)

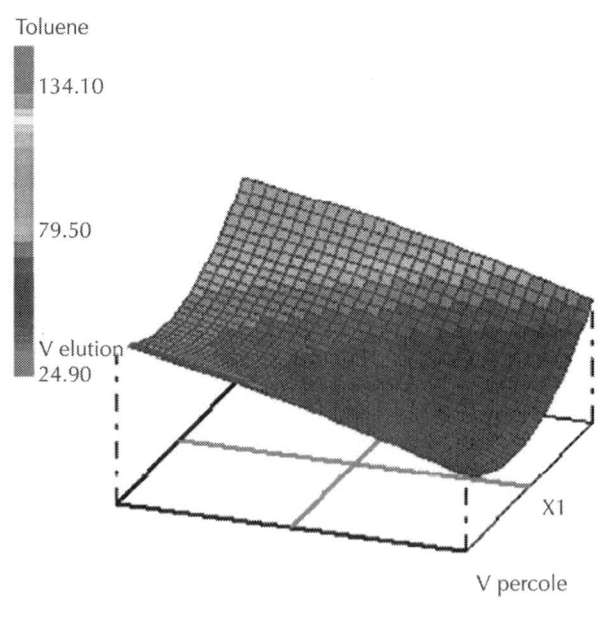

(b)

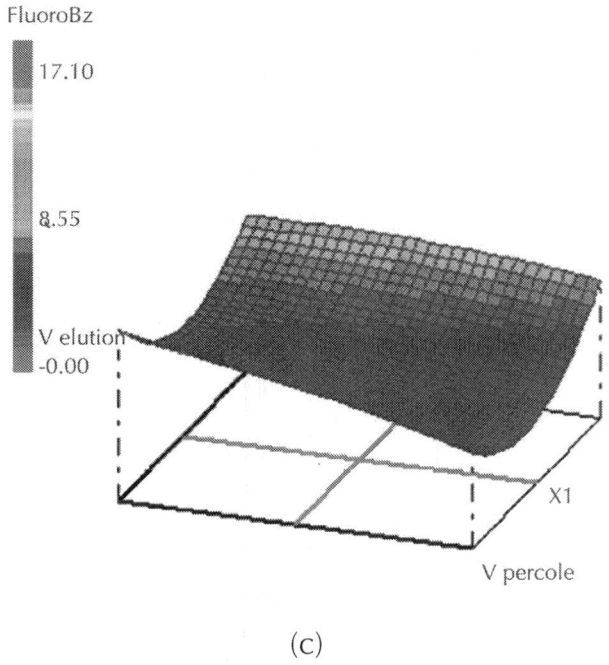

(c)

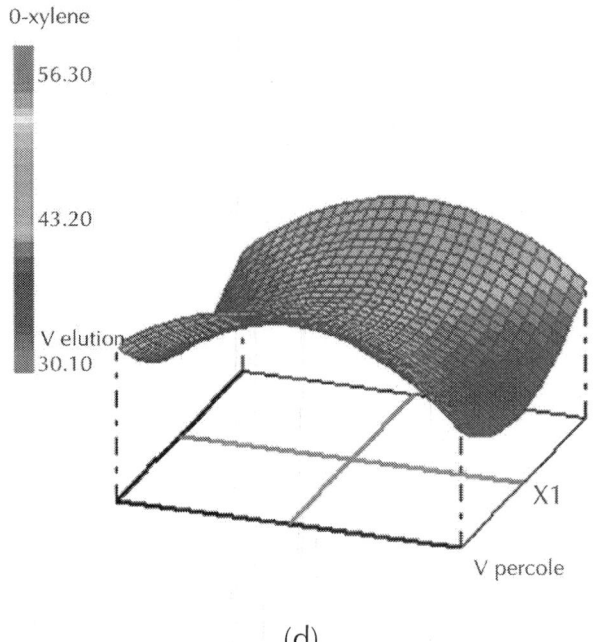

(d)

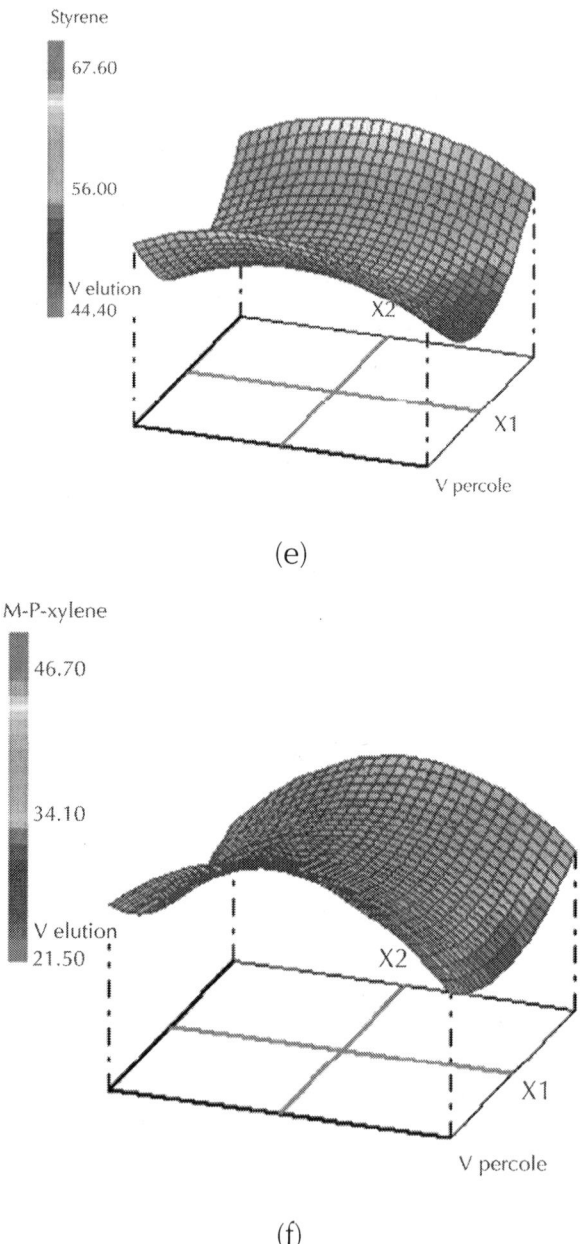

Figure 2: Response surfaces obtained by SPE-GC/FID of (a) fluoroben-zene; (b) toluene; (c) ethylbenezene; (d) m/p-xylene; (e) styrene; (f) o-xylene.

Table 3: Effect of the type of coating on the repeatability of the HS-SPME-GC/FID (n = 3): mixture of VOCs standards at 10 $\mu g \cdot L^{-1}$ for each compound, extraction (10 min at 25°C), salinity of 0.2 $g \cdot mL^{-1}$, headspace volume (10/20 mL)

Fiber coating	RSD%		
	PDMS	**PDMS/ DVB**	**CAR/ PDMS**
Benzene	35	15	77
Toluene	35	11	13
Chlorobenzene	29	15	21
Ethylbenzene	28	16	12
m/p-Xylene	25	9	20
Styrene	31	12	15
o-Xylene	22	15	26
Benzaldehyde	18	15	20

Optimization of the SPE

Global desirability of SPE (DSPE = 0.9563) value was satisfactory [42,46,47]. SPE extraction efficiency was very poor for benzene, fluorobenzene and benzaldehyde which makes difficult to reach acceptable sensitivity and repeatability of the analytical method. These analytes were thus not included in the SPE optimization. The four SPE studied factors were: the sample volume, the elution volume of acetonitrile, the keeper volume in elution phase, and the percentage of the evaporation of the extract volume under nitrogen stream. The minimum and maximum values of each factor were set after preliminary experiments. The studied range of the sample volume was set between 250 and 450 mL to ensure high load of analytes within a reasonable time. The optimum sample volume was 330 mL, since higher volumes require not only more loading time, but lead also to lower yields due to: 1) an increase of the VOCs loss by evaporation during this step and 2) a higher breakthrough of the analytes from the cartridges [48]. The trapped analytes were eluted under optimum conditions: with 10 mL of acetonitrile containing 108 µL of dodecane. Sample concentration was then carried out under a stream of nitrogen until 50% of evaporation of the extract volume. The efficiency of the acetonitrile volume on the SPE yields was tested ranging from 10 to 18 mL of the eluent. Despite that high acetonitrile volume can elute any residual trapped analytes from the cartridge; the optimum volume was 10 mL (minimum volume) since highest volumes require additional time of evaporation under nitrogen stream in the subsequent step. The volatiles analytes are thus lost by evaporation at this final step, which should be kept at its minimum (50% of sample volume evaporation found as optimum). The optimum volume of the keeper in the eluent composition was 108 µL, since enough dodecane should be added to reduce the evaporation of the analytes during and after elution. However, higher keeper volume decreases the analytes solubility in the acetonitrile phase and consequently decreases the elution efficiency.

Methods Validation

HS-SPME-GC/FID Validation

The HS-SPME-GC/FID was valid for all the studied analytes, with good linearity expressed as r^2(>0.988), in the range between LOQ and 100 $\mu g \cdot L^{-1}$, fulfilling the requirements of the AFNOR NF T 90 - 210 (May 2009). The precision of the experimental procedures was assessed at five concentration levels of analytes within the linearity range (5 replicates). The results of the HS-SPME showed good intermediate repeatability with relative standard deviations (RSDr) less than 15% for all the analytes, within the acceptance criteria (Table 4) [49].

Method accuracy was determined by considering both systematic and random errors. Accuracy was thus estimated from the uncertainty measurement of the analytical assay determined from the validation data according the LGC/VAM protocol and the ISO/DTS 21748 guide [49]. The uncertainty measurement of the HS-SPME was ranging from 0.02 to 15.83.

The limits of quantification are the lowest concentration of analytes that can be determined These quantitatively with an acceptable level of precision. limits were validated following the ANOR XP T 90-210 recommendations and their values were with HS-SPME far lower than MCLs, allowing the determination of these contaminants in bottled drinking water, with respect to the US EPA and WHO regulations (Table 4).

The method recoveries were determined using spiked water samples that were extracted and analyzed by GC/ FID under the optimized conditions. The recovery values obtained ranging from 90% - 117% for all the analytes by HS-SPME in the concentration range between LOQ and 100 $\mu g \cdot L^{-1}$ are very satisfactory.

SPE-GC/FID Validation

SPE-GC/FID validation results are summarized in Table 4. This method was valid following AFNOR NF T 90- 210 (May 2009) criteria for only four analytes (ethylbenzene, m/p-xylenes, styrene, and o-xylene) with $r^2 > 0.998$ in the range between LOQ and 250 $\mu g \cdot L^{-1}$. The intermediate repeatability of the SPE was also acceptable for the four valid analytes with RSDr less than 16%. The lowest uncertainty measurement of the SPE was 3.60 for styrene (Table 4).

SPE does not provide enough sensitivity for most of

Table 4: Validation of the analytical methods (AFNOR NF T 90 - 210) for the VOCs in water and their MCLs

	Analyte	LOQ (µg·L⁻¹)	Linearity range (µg·L⁻¹)	r² (n = 25)	RSD, % (n = 5)	Recovery range %	Measurement uncertainty	MCL (µg L⁻¹) US-EPA	MCL (µg L⁻¹) WHO
HS-SPME	Benzene	0.5	0.51 - 100	0.9989	0.6 - 14	97 - 117	0.31 - 1.47	5	10
	Toluene	1.1	1.1 - 100	0.9981	2 - 9	95 - 109	0.06 - 7.72	1000	700
	Chlorobenzene	0.5	0.05 - 100	0.9880	3 - 14	97 - 109	0.03 - 9.99	.	.
	Ethylbenzene	1.5	1.5 - 100	0.9980	2 - 10	90 - 105	0.02 - 4.41	700	300
	m/p-Xylenes	1.6	1.6 - 100	0.9885	2 - 15	94 - 111	0.03 - 11.91	10.000	500
	Styrene	0.2	0.17 - 100	0.9960	1 - 7	95 - 110	1.35 - 15.83	100	20
	o-Xylene	0.3	0.28 - 100	0.9932	3 - 12	90 - 115	1.31 - 9.33	10.000	500
	Benzaldehyde	1.2	1.22 - 100	0.9984	3 - 5	93 - 113	0.81 - 4.72	.	.
SPEᵃ	Ethylbenzene	70	70 - 250	0.9984	6 - 12	85 - 117	9.54 - 13.08	700	300
	m/p-Xylenes	70	70 - 250	0.9983	6 - 14	87 - 116	5.04 - 15.20	10.000	500
	Styrene	50	50 - 250	0.9989	4 - 9	90 - 115	3.60 - 13.71	100	20
	o-Xylene	50	50 - 250	0.9991	6 - 16	88 - 117	5.39 - 9.80	10.000	500

[a]The SPE-GC/FID was not valid for benzene, toluene, chlorobenzene, and benzaldehyde.

the analytes. LOQ was higher than MCL for styrene among the four valid analytes, following only USEPA standards (not WHO). SPE provides wider range of recoveries than HS-SPME ranging from 85% to 117% for the valid four analytes and for higher concentration levels (≤50/70 - 250 µg·L⁻¹≥) (Table 4).

When comparing the validation criteria of both methods and in the response to the essential need of high sensitivity to monitor traces of VOCs in bottled water, it is obvious that the SPE is less preferment than HS-SPME to be applied for VOCs screening in different bottled waters and to assess the migration of these molecules from packaging to the bottled water under several conditions of storage.

VOCs Background in the Blanks

Blank samples were analyzed to determine the VOCs concentrations as background levels that should be taken into account for LOQs determination as well for recoveries calculation. The main problem was that VOCs are found in blanks of ultra-pure water under our laboratory conditions. Blank chromatographic injections were made to ensure the absence of VOCs in the apparatus. Their detection in the blank chromatograms may be attributed to their presence at low levels in the ultra-pure water or due to contamination during the analytical procedural stages.

With SPE, traces of ethylbenzene, xylenes, and styrene were found in the blanks with the styrene found at the highest peak areas. With SPME, traces of almost all of the VOCs were found in the blanks with highest peak areas for toluene and m/p-xylenes. However, all compounds are, when found in the blanks, at very low amounts and not quantified. Their peak areas were subtracted from the corresponding ones in real samples.

VOCs Screening

A total of 43 different PET-bottled water samples were analyzed along with 11 PC-bottled water samples to assess the contamination of waters with VOCs. Glassbottled water (n = 2) was taken as reference, and comparison was carried out with tap waters (n = 6) to assess the global environmental contamination of drinking waters with the target analytes. The results of the HS-SPME analysis (Table 5) showed no contamination with all the analytes using the glass as packaging material. Analyzed tap water samples were contaminated only by toluene and m/p-xylenes below the LOQ and therefore far below their MCL, showing the safety of the Lebanese tap waters for human health with regards to VOCs. In plastic-bottled waters, benzene, chlorobenzene, styrene and benzaldehyde were not detected in all the samples whatever the plastic packaging is (PET or PC). Toluene was detected in all the samples but all concentrations were below the LOQ, similar to tap waters; thus the HS-SPME results confirm the presence of toluene in bottled waters in the same manner as obtained previously [25,50]. The source of contamination with toluene cannot be well established, since this pollutant is widely detected in the environmental waters, and all the concentrations in bottled waters are below the LOQ making difficult the study of migration under realistic conditions.

Ethylbenzene was detected in 13 over 43 PET-bottled waters and in 10 over 11 PC-bottled ones; whereas oxylene was detected in 4 of the PET-, and in 9 of the PCbottled waters. Ethylbenzene as well as toluene, benzene, and styrene are aromatic hydrocarbons that are commonly identified in PET samples subjected to temperatures between 200°C and 300°C due to thermal degradation [3,51]. However, in our study, no correlation was found between the detection of the aforementioned four VOCs, since when ethylbenzene and toluene are detected in PET-bottled waters, no traces of either benzene, or styrene are detected. The high volatility of benzene might be one of the reasons that cause its loss from bottled waters, either by diffusion through the plastic pores, or by volatilization at the direct opening of the bottle. m/pxylenes are detected at concentrations below their LOQ in all PETand PC-bottled waters. P-xylene is considered as residue from the production of tetrephtalique acid or its ester, the dimethylterephtalate (raw materials required to synthesis PET). Similarly, this VOC is part of not intentionally added substances (NIAS) that result often from PET degradation during its manufacturing process [52].

Even if some bottled waters were contaminated by the studied molecules, all the concentrations were far below their MCLs, showing that the Lebanese bottled waters are safe for human consumption close to tap waters, in respect to the VOCs contamination. However, the occurrence of the aforementioned contaminants in tap waters deserves future and continuous monitoring of the unpacked waters to assess any increase in their concentrations that could be harmful for public health.

VOCs Migration Study

In order to assess the influence of storage conditions of bottled waters on the VOCs migration from the PET material to the mineral water, the total of 20 PET-bottled waters were analyzed after their storage during 5 months, either at room temperature in dark, or under daily sunlight. Glass bottled waters were taken as reference and were stored with the aforementioned samples, under same conditions. Benzene, cholorobenzene, and benzaldehyde were not detected in all the analyzed samples (Table 6). Toluene was found in all the samples (<="" p=""> Styrene detection in samples after solar exposure cannot be attributed to migration since this compound remains below

Table 5: Concentrations range of VOCs ($\mu g \cdot L^{-1}$) in the water samples by HS-SPME-GC/FID

Packaging	PET						With-out	Glass	PC
Bottle size (L)	0.33	0.5	1	1.5	2	6	Tap water	0.75	18.9
Number of replicates	3	15	2	6	12	5	6	2	11
Benzene	n.d.	n.d.	n.d.	n.d.	n.d.	n.d.	n.d.	n.d.	n.d.
Toluene	<LOQ	<LOQ	<LOQ	<LOQ	<LOQ	<LOQ	<LOQ	n.d.	<LOQ
Chloroben-zene	n.d.	n.d.	n.d.	n.d.	n.d.	n.d.	n.d.	n.d.	n.d.
Ethylben-zene	<LOQ	<LOQ	n.d.	n.d.	n.d	<LOQ	n.d.	n.d.	n.d.- < LOQ
m/p-Xylene	<LOQ	<LOQ	<LOQ	<LOQ	<LOQ	<LOQ	<LOQ	n.d.	<LOQ
Styrene	n.d.	n.d.	n.d.	n.d.	n.d.	n.d.	n.d.	n.d.	n.d.
o-Xylene	<LOQ	n.d.	n.d.	n.d.	n.d.	<LOQ - 0.93	n.d.	n.d.	n.d.
Benzalde-hyde	n.d.	n.d.	n.d.	n.d.	n.d.	n.d.	n.d.	n.d.	n.d.

Table 6: VOCs concentrations (μg·L^{-1}) in bottled waters depending on the storage conditions

	Solar sunlight						Dark		
	PET			Glass			PET	Glass	
Size (L)	1	1.5	2	0.75	1	2	1.5	2	0.75
Number of replicates	2	3	5	1	2	5	3	5	1
Benzene	n.d.	n.d.	n.d.	n.d.	n.d.	n.d.	n.d.	n.d.	n.d.
Toluene	<LOQ	<LOQ	<LOQ	<LOQ	<LOQ	<LOQ	<LOQ	<LOQ	<LOQ
Chlorobenzene	n.d.	n.d.	n.d.	n.d.	n.d.	n.d.	n.d.	n.d.	n.d.
Ethylbenzene	<LOQ-2	<LOQ-1.8	<LOQ	n.d	n.d	<LOQ	<LOQ	<LOQ	n.d.
m/p-Xylene	<LOQ-2.2	<LOQ-2.2	<LOQ	n.d	<LOQ	<LOQ	<LOQ	<LOQ	n.d.
Styrene	<LOQ	<LOQ	n.d.	n.d.	n.d.	n.d.	n.d.	n.d.	n.d.
o-Xylene	<LOQ-1.1	<LOQ-1	<LOQ	n.d.	n.d.	<LOQ	<LOQ	<LOQ	n.d.
Benz-aldehyde	n.d.	n.d.	n.d.	n.d.	n.d.	n.d.	n.d.	n.d.	n.d.

LOQ. Sunlight seems to promote only ethylbenzene and xylenes migration from PET bottles, since quantifiable concentrations were determined after 5-months of samples exposure to sunlight. However, the concentrations remain far lower than the MCLs, with no considerable risks on human health. These contaminants have been previously quantified in other bottled waters at higher concentrations [25].

According to the Commission Directive 2002/72/EC, the safety of plastic materials such as PET intended to come into contact with foodstuffs (i.e. mineral water) should be assessed by means of specific migration tests. For this purpose, a total of 10 samples were analyzed freshly after collection and 10 similar samples were incubated at 40°C during 10 days then analyzed and the results were compared. Extended period of exposure until 20 days was also studied. The results in Table 7 showed no significant and quantifiable increase in the migration of the VOCs after 10 and 20 days of storage at 40°C.

The o-xylene quantified at an average of 0.9 $\mu g \cdot L^{-1}$ at day 10, is expected to reach 5.4 $\mu g \cdot L^{-1}$(when multiplied by a factor of six following Crank equation), after 1 year of usual storage (2002/72/EC), but still below its MCL.

Similar safety of the Lebanese PET used to bottle waters in respect to VOCs migration was demonstrated in a previous article with Head space trap GC/FID [50].

CONCLUSIONS

The HS-SPME-GC/FID was valid (Afnor NFT 90-210

Table 7: Assessment of the migration of VOCs from PET to bottled waters when stored in the oven at 40°C

Concentrations range ($\mu g \cdot L-1$) - n = 10			
Incubation time	Day 0	Day 10	Day 20

Benzene	n.d.	n.d.	n.d.
Toluene	<LOQ	<LOQ	<LOQ
Chlorobenzene	n.d.	n.d.	n.d.
Ethylbenzene	<LOQ	<LOQ	<LOQ
m/p-Xylene	<LOQ	<LOQ	<LOQ
Styrene	n.d.	n.d.	n.d.
o-Xylene	<LOQ – 0.93	<LOQ – 0.89	<LOQ – 0.91
Benzaldehyde	n.d.	n.d.	n.d.

(May 2009)) under optimized conditions: 0.2 $g \cdot mL^{-1}$ of NaCl; extraction during 31 min at 25°C; Headspace vol ume of 10 mL in a 20 mL vial, to analyze the studied VOCs with LOQs ranging from 0.2 (styrene) to 1.6 $\mu g \cdot L^{-1}$ (m/p-xylenes), which are far lower than their MCLs. SPE-GC/FID showed poorer performances, even after optimization using CCD, with LOQ above 50 $\mu g \cdot L^{-1}$ for the validated molecules (4/8 VOCs). All analyzed samples were not contaminated with benzene, chlorobenzene, styrene and benzaldehyde, similarly to tap waters. The contamination with the other VOCs when occurs was at very low concentration levels, below the LOQ. The safety of the Lebanese drinking waters is thus proved in respect to VOCs analysis. Analysis performed on PETbottled waters stored at 40°C for 10 and 20 consecutive days did not show any significant VOCs migration and the packaging can be considered safe at its expiry date (Directive 2002/72/EC). The exposure to sunlight during 5 months seems to promote the migration of ethylbenzene and xylenes, but their concentrations in bottled waters remain below the MCLs.

ACKNOWLEDGEMENTS

The authors are grateful to the EDST of the Lebanese University and to the Lebanese National Council for Scientific Research (CNRS) for the financial support of this project.

REFERENCES

1. B. Nijssen, T. Kamperman and J. Jetten, "Acetaldehyde in Mineral Water Stored in Polyethylene Terephthalate (PET) Bottles: Odour Threshold and Quantification," Packaging Technology and Science, Vol. 9, No. 4, 1996, pp. 175-185.

2. S. V. Leivadara, A. D. Nikolaou and T. D. Lekkas, "Determination of Organic Compounds in Bottled Waters," Food Chemistry, Vol. 108, No. 1, 2008, pp. 277-286. doi:10.1016/j.foodchem.2007.10.031

3. C. Bach, X. Dauchy, L. David and S. Etienne, "Chemical Compounds and Toxicological Assessments of Drinking Water Stored in Polyethylene Terephthalate (PET) Bottles: A Source of Controversy Reviewed," Water Research, Vol. 46, No. 3, 2012, pp. 571-583.doi.:10.1016/j.watres.2011.11.062

4. Ikem, "Measurement of Volatile Organic Compounds in Bottled and Tap Waters by Purge and Trap GC-MS: Are Drinking Water Types Different?" Journal of Food Composition and Analysis, Vol. 23, No. 1, 2010, pp. 70- 77. doi:10.1016/j.jfca.2009.05.005

5. M. Farhadian, C. Vachelard, D. Duchez and C. Larroche, "In Situ Bioremediation of Monoaromatic Pollutants in Ground Water: A Review," Bioresource Technology, Vol. 99, No. 13, 2008, pp. 5296-5308. doi:10.1016/j.biortech.2007.10.025

6. M. E. Mahmoud and H. M. AlBishri, "Identification of Dissolved Organic Species in Non-Drinking Tap Water by Solid-Phase Extraction and Gas Chromatography-Mass Spectrometry," Journal of Saudi Chemical Society, Vol. 14, No. 1, 2010, pp. 105-115.doi:10.1016/j.jscs.2009.12.017

7. S. Fakirov, "Handbook of Thermoplastic Polyesters," Wiley-VCH, Weinheim, 2002.doi:10.1002/3527601961

8. P. Westerhoff, P. Prapaipong, E. Shock and A. Hillaireau, "Antimony Leaching from Polyethylene Terephthalate (PET) Plastic Used for Bottled Drinking Water," Water

Research, Vol. 42, No. 3, 2008, pp. 551-556. doi:10.1016/j. watres.2007.07.048

9.   C. Bach, X. Dauchy, L. David and S. Etienne, "PhysicoChemical Study of PET Bottles and PET Bottled Water," Matériaux & Techniques, Vol. 99, No. 4, 2011, pp. 391- 408.doi:10.1051/ mattech/2011006

10. S.-H. Nam, Y.-M. Seo and M.-G. Kim, "Bisphenol A Migration from Polycarbonate Baby Bottle with Repeated Use," Chemosphere, Vol. 79, No. 9, 2010, pp. 949-952. doi:10.1016/j.chemosphere.2010.02.049

11. H. H. Le, E. M. Carlson, J. Chua, P. Belcher and M. Scott, "Bisphenol A Is Released from Polycarbonate Drinking Bottles and Mimics the Neurotoxic Actions of Estrogen in Developing Cerebellar Neurons," Toxicology Letters, Vol. 176, No. 2, 2008, pp. 149-156.doi:10.1016/j.toxlet.2007.11.001

12. J. Ji, C. Deng, W. Shen and X. Zhang, "Field Analysis of Benzene, Toluene, Ethylbenzene and Xylene in Water by Portable Gas Chromatography-Microflame Ionization Detector Combined with Headspace Solid-Phase Microextraction," Talanta, Vol. 69, No. 4, 2006, pp. 894-899. doi:10.1016/j. talanta.2005.11.032

13. D. Li, B. Han, Z. Liu and D. Zhao, "Phase Behavior of Supercritical CO_2/Styrene/Poly(ethylene Terephthlate) (PET) System and Preparation of Polystyrene/PET Composites," Polymer, Vol. 42, No. 6, 2001, pp. 2331-2337.doi:10.1016/ S0032-3861(00)00601-7

14. M. Dzieciol and J. Trzeszczynski, "Studies of Temperature Influence on Volatile Thermal Degradation Products of Poly(ethylene terephthalate)," Journal of Applied Polymer Science, Vol. 69, No. 12, 1998, pp. 2377-2381. doi:10.1002/(SICI)1097-4628(19980919)69:12<2377::AID-APP9>3.0.CO;2-5

15. R. Di Felice, D. Cazzola, S. Cobror and L. Oriani, "Oxygen Permeation in PET Bottles with Passive and Active Walls," Packaging Technology and Science, Vol. 21, No. 7, 2008, pp.

405-415. doi:10.1002/pts.820

16. K. E. Ozlem, "Acetaldehyde Migration from Polyethylene Terephthalate Bottles into Carbonated Beverages in Turkiye," International Journal of Food Science and Technology, Vol. 43, No. 2, 2008, pp. 333-338. doi:10.1111/j.1365-2621.2006.01443.x

17. C. Dutra, P. D. Freire, M. T. de Alvarenga, N. C. Reyes and F. G. Reyes, "Determination of Volatile Organic Compounds in Recycled Polyethylene Terephthalate and HighDensity Polyethylene by Headspace Solid Phase Microextraction Gas Chromatography Mass Spectrometry to Evaluate the Efficiency of Recycling Processes," Journal of Chromatography A, Vol. 1218, No. 10, 2011, pp. 1319- 1330.doi:10.1016/j.chroma.2010.12.099

18. S. Fabris, M. T. Freire, W. de Alvarenga, R. Reyes and F. Guillermo, "A Method to Determine Volatile Contaminants in Polyethylene Terephthalate (PET) Packages by HDC-GC-FID and its Application to Post-Consumer Materials," Ciência e Tecnologia de Alimentos, Vol. 30, No. 4, 2010, pp. 1046-1055. doi:10.1590/S0101-20612010000400033

19. F. L. Bayer, "Polyethylene Terephthalate Recycling for FoodContact Applications: Testing, Safety and Technologies: a Global Perspective," Food Additives and Contaminants, Vol. 19, No. S1, 2002, pp. 111-134. doi:10.1080/02652030110083694

20. Lara-Gonzalo, J. E. Sánchez-Uría, E. Segovia-García and A. Sanz-Medel, "Critical Comparison of Automated Purge and Trap and Solid-Phase Microextraction for Routine Determination of Volatile Organic Compounds in Drinking Waters by GC-MS," Talanta, Vol. 74, No. 5, 2008, pp. 1455-1462. doi:10.1016/j.talanta.2007.09.036

21. E. C. Commission Regulation (EU), "No 10/2011 of 14 January 2011 on Plastic Materials and Articles Intended to Come in Contact with Food," Official Journal of the European Commission, 2011, pp. L12-L89.

22.   US-EPA, "National Primary Drinking Water Regulations," 2009. http://www.epa.gov/ogwdw/consumer/pdf/mcl.pdf

23.   W. H. Organization, "Guidelines for Drinking-Water Quality," 2008.   http://www.who.int/water_sanitation_health/dwq/ GDW8rev1and2.pdf

24.   US-EPA, "Measurement of Purgeable Organic Compounds in Water by Capillary Column Gas Chromatography/ Mass Spectrometry," 1992.

25.   H. F. Al-Mudhaf, F. A. Alsharifi and A.-S. I. Abu-Shady, "A Survey of Organic Contaminants in Household and Bottled Drinking Waters in Kuwait," Science of the Total Environment, Vol. 407, No. 5, 2009, pp. 1658-1668.doi:10.1016/j. scitotenv.2008.10.057

26.   H. S. Dórea, J. R. L. Bispo, A. A. S. Kennedy, B. B. Cunha, S. Navickiene, J. P. H. Alves, L P. C. Romão and C. A. B. Garcia, "Analysis of BTEX, PAHs and Metals in the Oilfield Produced Water in the State of Sergipe, Brazil," Microchemical Journal, Vol. 85, No. 2, 2007, pp. 234- 238. doi:10.1016/j. microc.2006.06.002

27.   P. Kavcar, M. Odabasi, M. Kitis, F. Inal and S. C. Sofuoglu, "Occurrence, Oral Exposure and Risk Assessment of Volatile Organic Compounds in Drinking Water for İzmir," Water Research, Vol. 40, No. 17, 2006, pp. 3219-3230. doi:10.1016/j.watres.2006.07.002

28.   R. M. Cavalcante, M. V. F. de Andrade, R. V. Marins and L. D. M. Oliveira, "Development of a Headspace-Gas Chromatography (HS-GC-PID-FID) Method for the Determination of VOCs in Environmental Aqueous Matrices: Optimization, Verification and Elimination of Matrix Effect and VOC Distribution on the Fortaleza Coast, Brazil," Microchemical Journal, Vol. 92, No. 2, 2010, pp. 337- 343. doi:10.1016/j.microc.2010.05.014

29.   K. E. F. Sieg and W. Püttmann, "Analysis of Benzene, Toluene, Ethylbenzene, Xylenes and n-Aldehydes in Melted Snow Water via Solid-Phase Dynamic Extraction Combined with Gas Chromatography/Mass Spectrometry," Journal of

Chromatography A, Vol. 1178, No. 1-2, 2008, pp. 178-186. doi:10.1016/j.chroma.2007.11.025

30. C. Aeppli, M. Berg, T. B. Hofstetter, R. Kipfer and R. P. Schwarzenbach, "Simultaneous Quantification of Polar and non-Polar Volatile Organic Compounds in Water Samples by Direct Aqueous Injection-Gas Chromatography/ Mass Spectrometry," Journal of Chromatography A, Vol. 118, No. 1-2, 2008, pp. 116-124.doi:10.1016/j.chroma.2007.12.043

31. H. Jurdáková, A. Kraus, W. Lorenz, R. Kubinec, Ž. Krkošová, J. Blaško, I. Ostrovský, L. Soják and V. Pacáková, "Determination of Gasoline and BTEX in Water Samples by Gas Chromatography with Direct Aqueous Injection," Petroleum & Coal, Vol. 47, No. 3, 2005, pp. 49-53.

32. M. A. Hossain, M. J. Kabir and S. M. Salehuddin, "Determination of Toxic Toluene, Xylene and Cumene in Different Lake Waters," International Journal of Environmental Research, Vol. 4, No. 2, 2010, pp. 341-346.

33. M. K. Meney and C. M. Davidson, "Use of Solid-Phase Extraction in the Determination of Benzene, Toluene, Ethylbenzene, Xylene and Cumene in Spiked Soil and Investigation of Soil Spiking Methods," Analyst, Vol. 123, No. 2, 1998, pp. 195-200.doi:10.1039/a706258c

34. Serrano and M. Gallego, "Fullerenes as Sorbent Materials for Benzene, Toluene, Ethylbenzene, and Xylene Isomers Preconcentration," Journal of Separation Science, Vol. 29, No. 1, 2006, pp. 33-40. doi:10.1002/jssc.200500200

35. S. Nakamura and S. Daishima, "Simultaneous Determination of 22 Volatile Organic Compounds, Methyltertbutyl Ether, 1,4-Dioxane, 2-Methylisoborneol and Geosmin in Water by Headspace Solid Phase MicroextractionGas Chromatography-Mass Spectrometry," Analytica Chimica Acta, Vol. 548, No. 1-2, 2005, pp. 79-85.doi:10.1016/j.aca.2005.05.077

36. Arambarri, M. Lasa, R. Garcia and E. Millán, "Determination of Fuel Dialkyl Ethers and BTEX in Water using Headspace Solid-Phase Microextraction and Gas Chromatography-

Flame Ionization Detection," Journal of Chromatography A, Vol. 1033, No. 2, 2004, pp. 193- 203. doi:10.1016/j. chroma.2004.01.046

37. V. H. Niri, L. Bragg and J. Pawliszyn, "Fast Analysis of Volatile Organic Compounds and Disinfection by-Products in Drinking Water using Solid-Phase Microextraction—Gas Chromatography/Time-of-Flight Mass Spectrometry," Journal of Chromatography A, Vol. 1201, No. 2, 2008, pp. 222-227. doi:10.1016/j.chroma.2008.03.062

38. T. C. Schmidt, B. H. Stefan, P. Rolf and R. Forster, "Occurrence and Fate Modeling of MTBE and BTEX Compounds in a Swiss Lake used as Drinking Water Supply," Water Research, Vol. 38, No. 6, 2004, pp. 1520-1529. doi:10.1016/j. watres.2003.12.027

39. C. M. Almeida and L. V. Boas, "Analysis of BTEX and Other Substituted Benzenes in Water using Headspace SPME-GC-FID: Method Validation," Journal of Environmental Monitoring, Vol. 6, No.1, 2004, pp. 80-88. doi:10.1039/b307053k

40. Gaujac, S. E. Elissandro, N. Sandro, S. L. C. Ferreira and S. D. Haroldo, "Multivariate Optimization of a Solid Phase Microextraction-Headspace Procedure for the Determination of Benzene, Toluene, Ethylbenzene and Xylenes in Effluent Samples from a Waste Treatment Plant," Journal of Chromatography A, Vol. 1203, No. 1, 2008, pp. 99-104. doi:10.1016/j.chroma.2008.06.022

41. C. Flórez Menéndez, M. L. Fernández Sánchez, J. E. Sánchez Uría, E. Fernández Martínez and A. Sanz-Medel. "Static Headspace, Solid-Phase Microextraction and Headspace Solid-Phase Microextraction for BTEX Determination in Aqueous Samples by Gas Chromatography," Analytica Chimica Acta, Vol. 415, No. 1-2, 2000, pp. 9-20.doi:10.1016/S0003-2670(00)00862-X

42. F. Bianchi, M. Careri, E. Marengo and M. Musci, "Use of Experimental Design for the Purge-and-Trap-Gas

Chromatography–Mass Spectrometry Determination of Methyl Tert.-Butyl Ether, Tert.-Butyl Alcohol and BTEX in Groundwater at Trace Level," Journal of Chromatography A, Vol. 975, No. 1, 2002, pp. 113-121. doi:10.1016/S0021-9673(02)00881-6

43. Water Quality, "Protocol for the Initial Method Performance Assessment in a Laboratory," AF-NOR NF T90-210, 2009.

44. Neter, M. H. Kutner, C. J. Nachtsheim and W. Wasserman, "Applied Linear Statistical Models," McGrawHill, Chicago, 1998.

45. M.-R. Lee, C.-M. Chang and J. Dou, "Determination of Benzene, Toluene, Ethylbenzene, Xylenes in Water at sub-ngl^{-1} Levels by Solid-Phase Microextraction Coupled to Cryo-Trap Gas Chromatography—Mass Spectrometry," Chemosphere, Vol. 69, No. 9, 2007, pp. 1381-1387. doi:10.1016/j. chemosphere.2007.05.004

46. J. Wang and W. Wan, "Application of Desirability Function Based on Neural Network for Optimizing Bio-Hydrogen Production Process," International Journal of Hydrogen Energy, Vol. 34, No. 3, 2009, pp. 1253-1259. doi:10.1016/j. ijhydene.2008.11.055

47. H. H. Nguyen, N. Jang and S. H. Choi, "Multiresponse Optimization Based on the Desirability Function for a Pervaporation Process for Producing Anhydrous Ethanol," Korean Journal of Chemical Engineering, Vol. 26, No. 1, 2009, pp. 1-6.doi:10.1007/s11814-009-0001-5

48. E. M. Thurman and M. S. Mills, "Solid Phase Extraction: Principles and Practice," Wiley, New York, 1998.

49. G. González and M. Ángeles Herrador, "A Practical Guide to Analytical Method Validation, Including Measurement Uncertainty and Accuracy Profiles," TrAC Trends in Analytical Chemistry, Vol. 26, No. 3, 2007, pp. 227- 238. doi:10.1016/j. trac.2007.01.009

50. Al Rayes, C. O. Saliba, A. Ghanem and J. Randon, "BTES and Aldehydes Analysis in PET-Bottled Water in Lebanon," Food

Additives and Contaminants: Part B: Surveillance, Vol. 5, No. 3, 2012, pp. 221-227. doi:10.1080/19393210.2012.698311

51. Dzie͵cioł and J. Trzeszczynski, "Volatile Products of Poly(Ethylene Terephthalate) Thermal Degradation in Nitrogen Atmosphere," Journal of Applied Polymer Science, Vol. 77, No. 9, 2000, pp. 1894-1901. doi:10.1002/1097-4628(20000829)77:9<1894::AID-APP5>3.0.CO;2-Y

52. F. Alario, C. Marcilly and M. Barraqué, "BTX: Benzène, Toluène, Xylène," Techniques de l'ingénieur, 2005, 4p.

Chapter 3

Geochemical Modeling of Trivalent Chromium Migration in Saline-Sodic Soil during Lasagna Process: Impact on Soil Physicochemical Properties

Salihu Lukman[1], Alaadin Bukhari[2], Muhammad H. Al-Malack[2], N uhu D. Mu'azu[3], and Mohammed H. Essa[2]

[1]Department of Civil Engineering, ACHB, King Fahd University of Petroleum and Minerals, Hafar Al-Batin 31991, Saudi Arabia

[2]Department of Environmental Engineering, University of Dammam, Dammam, Saudi Arabia

[3]Environmental Engineering Department, University of Dammam, Dammam 31451, Saudi Arabia

ABSTRACT

Trivalent Cr is one of the heavy metals that are difficult to be removed from soil using electrokinetic study because of its geochemical properties. High buffering capacity soil is expected to reduce the mobility of the trivalent Cr and subsequently reduce the remedial efficiency thereby complicating the remediation process. In this study, geochemical modeling and migration of trivalent Cr in saline-sodic soil (high buffering capacity and alkaline) during integrated electrokinetics-adsorption remediation, called the Lasagna process, were investigated. The remedial efficiency of trivalent Cr in addition to the impacts of the Lasagna process on the physicochemical properties of the soil was studied. Box-Behnken design was used to study the interaction effects of voltage gradient, initial contaminant concentration, and polarity reversal rate on the soil pH, electroosmotic volume, soil electrical conductivity, current, and remedial efficiency of trivalent Cr in saline-sodic soil that was artificially spiked with Cr, Cu, Cd, Pb, Hg, phenol, and kerosene. Overall desirability of 0.715 was attained at the following optimal conditions: voltage gradient 0.36 V/cm; polarity reversal rate 17.63 hr; soil pH 10.0. Under these conditions, the expected trivalent Cr remedial efficiency is 64.75 %.

INTRODUCTION

In early 1992, a discussion took place between the then Monsanto Chief Executive Officer (CEO) and Administrator of the United States Environmental Protection Agency (USEPA) which ultimately led to the invention of the Lasagna process [1]. In the late 1993, Brodsky and Ho of Monsanto filed the first Lasagna U.S. patent followed by a second one, all published in 1995 [2, 3]. In the Lasagna process, contaminated soil is remediated by creating at least one liquid permeable zone within a contaminated soil region and turning it into treatment zone. Appropriate materials (sorbents, catalytic agents, microbes, oxidants, and buffers) are then introduced into the treatment zone. An electrode is placed at the first end of the contaminated soil region and another of opposite charge is placed at the opposite end of the contaminated soil region. A direct electric current is then transmitted through the contaminated soil region between the two electrodes. This causes movement of water and dissolved organic and inorganic materials in subsurface soils from one electrode (anode) to the other (cathode) under electroosmosis as a result of current movement from anode to cathode. In 1802 electroosmosis was first observed; detailed study of the mechanism was done by Reuss [4] in his classic experiment reported by Abramson [5]. In 1909, Freundlich and Neumann [6] provided the general name "electrokinetic phenomena" to refer to the electrically driven mass flow of dissolved contaminants and pore fluid transport in soils induced by an applied DC voltage. It is made up of transport of pore fluid via electroosmosis (EO) and transport of ions or charged species via electromigration [7]. The direction and quantity of contaminant movement are influenced by the contaminant concentration, solubility, speciation, degree of hydrophobicity, soil type and structure, and the mobility of contaminant ions, as well as the interfacial chemistry and the conductivity of the soil pore fluid [8]. The remedial efficiency generally depends on the nature of the contaminants and soil properties, such as pH, permeability, adsorption capacity, buffering capacity, and geochemical processes (such as acid/base reactions and migration, dissolution/precipitation, redox reactions, complexation, and speciation) [7, 9]. Saline-sodic soils possess electrical conductivity above 4 dS/m, soil paste pH greater than 8.2, and exchangeable sodium percentage greater than 15 [10, 11].

First application of electrokinetics took place in India in the 1930s. It was used to remove excess salts from alkali soils in order to restore it to arable condition [13]. Following its invention in 1993, extensive studies started in 1994 in bench-scale [14] then scaled up in a pilot-scale under laboratory conditions. The first field test called "Phase I: Small Field Test" was conducted in 1995 at the Paducah gaseous diffusion plant (PGDP) site whose soil was contaminated with trichloroethylene (TCE). Full-scale remediation using the Lasagna technology was undertaken at two other contaminated sites in the United States [15]. The specific details of all the Lasagna process implementations are presented in Tables 1 and 2. It is noteworthy that only two of the studies have considered simultaneous removal of contaminant mixture. All others have dealt with only a single organic compound or heavy metal. It has already been observed that contaminated soils do not contain single contaminants only but usually several pollutants appear in the soil in mixed components [16–19].

Table 1: Applications of Lasagna process at bench-scales from inception to date

Treatment zone material	Contaminant	Soil type	Cell dimensions (length × width × depth)	Polarity reversal/ downtime	Removal efficiency, %	Voltage gradient, (V/cm)/ current (mA)	Power consumption, kWhr/m³	Run time, days	Electroosmotic conductivity, cm²V⁻¹s⁻¹ (×10⁻⁵)	Treatment zone spacing, cm	Reference
AC* + sand, bacteria + AC + sawdust	p-nitrophenol	Kaolinite	10 cm ID, 21.6 cm long	Yes/ continuous	90–99	1–7/3 (constant)	10	20	2.5	6	[14, 20]
AC (Bamboo charcoal)	Cd	Sandy loam	24 cm × 10 cm × 8 cm	Yes/ continuous	79.6	1/7–27	—	12	—	10	[21]
AC (Bamboo charcoal)	Cd	Kaolin	24 cm × 10 cm × 8 cm	No/ continuous	93	1/3–23	—	8	—	10	[22]
AC (Bamboo charcoal)	2,4-dichlorophenol and Cd	Sandy loam	24 cm × 10 cm × 10 cm	Yes/ continuous	75.97 (Cd); 54.92 (2,4-dichlorophenol)	1/Variable	121.91–128.48	10.5	—	16	[23]
GAC**	Cr, Cd, Cu, Pb, Hg, Zn, phenol, kerosene	Saline-sodic clay	24 cm × 10 cm × 12 cm	No/ continuous	75.9 (Cr); 34.4 (Cd); 41 (Cu); 55.8 (Pb); 92.49 (Hg); 26.8 (Zn); 100 (phenol); 49.8 (kerosene)	0.6–1/880	1777–4273	21	425	6	[24]

Table 2: Applications of Lasagna process at pilot- and field-scales from inception to date

Treatment zone material	Contaminant	Soil type	Site dimensions (length × width × depth)	Polarity reversal/ downtime	Removal efficiency, %	Voltage gradient, (V/cm)/Current (A)	Power consumption, kwh/m³	Run time, month	Electroosmotic conductivity, cm²V⁻¹s⁻¹ (×10⁻⁵)	Treatment zone spacing, cm	Reference
AC + sand	p-nitrophenol	Kaolin/ Kaolinite/ clay loam	1.22 m × 0.61 m × 0.61 m	Yes/ continuous	98	1 (constant)/96.2 (based on current density)	51	3	0.56–1.7	35.56	[20]
GAC¹	TCE²	Clay loam	4.6 m × 3 m × 4.6 m	Yes/ continuous	99	0.35–0.45/40 (constant)	—	4	1.2	60	[25]
Iron filings + kaolin	TCE	Clay loam	6.4 m × 9.2 m × 13.7 m	Yes/3-week	95–99	0.23–0.31 (constant)/ 110–200	—	12	1.2	60 & 150	[26]
Iron filings + kaolin	TCE	Clay loam	27.4 m × 22 m × 13.5 m	No/pulse mode	99	0.15–0.26 (constant)/ 500–700	—	24	—	150	[15, 27]
Iron filings + kaolin	TCE	Clay loam	33 m × 24 m × 7.5 m	60 (after 1 year)	—	0.16 (constant)/ 250–400	—	24	—	150	[15]

The geochemical properties of the most stable forms of Cr, that is, trivalent and hexavalent Cr under electrokinetic remediation, have been extensively studied in different types of soils (kaolin, glacial till, etc.) by Reddy and his coworkers and other investigators [28–36]. The trivalent Cr, though considered relatively nontoxic compared to the hexavalent Cr, exists in the subsurface environments as cation, Cr^{3+}, and in the following hydroxocomplex forms: $Cr(OH)_4^-$, $CrOH^{2+}$, and $Cr(OH)_3^0$. Cr^{3+} and $CrOH^{2+}$ ions are mostly prevalent at soil pH values less than 6, while $Cr(OH)_4^-$ and $Cr(OH)_3^0$ ions prevail when pH is greater than 11.8. The redox state also affects the Cr state with reduced state favoring the presence of the trivalent Cr while the oxidized state favors the existence of the hexavalent Cr. Most of the trivalent Cr species are less mobile because of their low solubility over wide pH range (<12) and may be readily adsorbed by the negatively charged clay surfaces. There exists redox state in the subsurface environment because of the generation of oxygen and hydrogen gases at the electrodes in addition to the possible presence of iron (reducing agent), manganese (oxidizing agent), or microorganisms. The redox potential (Eh) and soil pH determine the possible oxidation of Cr from the trivalent to hexavalent form as shown in Figure 1. Chinthamreddy and Reddy [28] have found no significant oxidation of trivalent Cr in high buffering capacity soil such as glacial till.

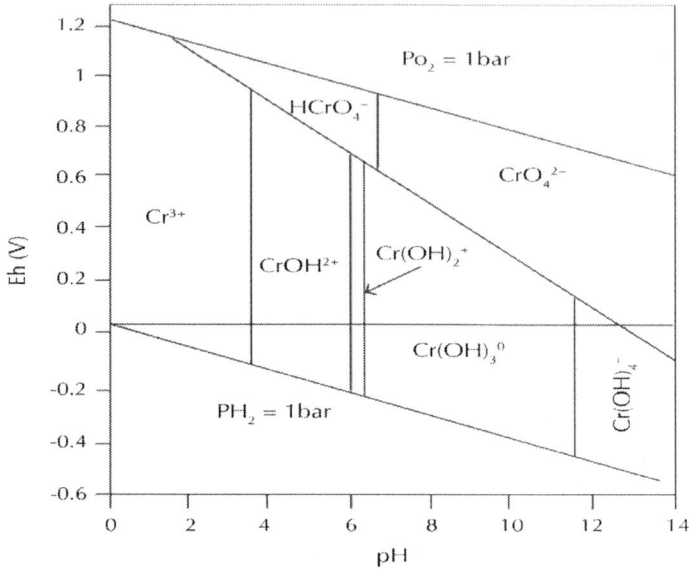

Figure 1: Redox potential (Eh)-pH diagram for Cr–O–H system [12].

Empirical modeling using response surface methodology (RSM) offers great and numerous advantages which include large amount of information from a small number of experiments, evaluation of simultaneous interaction effects of the independent parameters on the responses, and simultaneous optimization of multiple factors and responses for obtaining optimal conditions [37, 38]. The key success of RSM is uncovering interactions of factors which cannot be achieved using the traditional one-factor-at-a-time (OFAT) optimization approach [39]. Fundamental understanding of the physics and chemistry which governs the process is essential in determining the influential factors to be investigated and their levels or ranges are necessary for successful implementation of RSM for any process modeling and optimization. Basically, there exist four different experimental designs for RSM implementation: 3-level factorial design (3FD), Box-Behnken design (BBD), central composite design (CCD), and Doehlert design (DD). Bezerra et al. [38] have reviewed each of these design methods. The Box-Behnken design is obtained by combining two-level factorial designs with incomplete block designs followed by adding a specified number of replicated center points. BBD is preferred when investigating three (3) factors using RSM, because it will give enough information for analyzing factor-response interactions from the least experimental runs when compared to 3FD and CCD. Some microbially-driven biotransformation processes may affect the soil physicochemical properties after electrokinetic remediation because of the passage of electric current and development of pH gradients [40]. These lead to original soil mineral degradation and alteration via biotransformation. While biotransformation deals with the bioweathering and alteration or degradation of clay minerals, biomineralization refers to the formation of amorphous and crystalline materials from aqueous ions by biologically mediated processes. In addition to current and pH gradients, heavy metals also cause to affect the following biological assays: soil microbial biomass carbon, enzyme activity, basal soil respiration, and earthworm assays and seed assays [41–44]. Given the aforementioned intricacies of the geochemical behavior and migration of trivalent Cr in soil during electrokinetic remediation, this study was aimed at investigating trivalent Cr migration and remedial efficiency in high buffering capacity and alkaline soil during electrokinetic study in addition to the impacts of the soil remediation on the physicochemical properties of the soil. A carefully designed experiment using BBD was used to study the interaction effects of voltage gradient, initial contaminant concentration, and polarity reversal rate on the trivalent Cr remedial efficiency in saline-sodic soil that was artificially spiked with Cr, Cu, Cd, Pb, Hg, phenol, and kerosene using RSM modeling and optimization tools.

MATERIALS AND METHODS

Characterization

Natural saline-sodic clay, obtained from Al-Hassa Oasis, Saudi Arabia, was used in this study. The soil has the following characteristics: pH (8.3), moisture content (3.91%), soil organic matter (2.59%), electrical conductivity (15.24 dS/m), specific surface area (9.07 m²/g), pore volume (0.014 cm³/g), pore size (62.55 Å)—mineralogy from X-ray diffraction (XRD), quartz (SiO_2) (87.4%), calcite ($CaCO_3$) (5.2%), and dolomite ($CaMg(CO_3)_2$) (7.4%). X-ray fluorescence spectroscopy (XRF) revealed that the soil consists of the following elements: Ca (37.64%), Si (34.73%), Fe (10.41%), Al (7.6%), K (3.42%), Mg (2.48%), Pd (2.85%), and Ti (0.86%). These properties were determined using methods of the American Society of Testing and Materials (ASTM) standards and were reported elsewhere [45]. The granular activated carbon (GAC) used in the present study whose surface area is 952 m²/g was produced locally from date palm pits using phosphoric acid impregnation method. Its characterization and properties have been reported elsewhere [46, 47].

Adsorption Testing

Single and competitive adsorption of five heavy metals (Cr, Cd, Cu, Zn, and Pb) were performed to determine the selectivity sequence and to understand the adsorption behavior of these metals under different pH conditions. This is particularly important to this study, because soil mineralogy affects heavy metal adsorption behavior and selectivity sequence under different pH conditions. Lukman et al. [45] reported the detailed procedures carried out for the competitive adsorption testing.

Coupled Electrokinetics-Adsorption Study

Fifteen (15) bench-scale experiments, each having a 21-day run time, were designed and performed to investigate the migration and distribution of trivalent Cr in a contaminant mixture using the coupled electrokinetics-adsorption technique and to understand the operating variables' effects on saline-sodic soil.

Reactor Design and Experimental Procedures

The Plexiglas reactor total volume was about 2268 cm^3, made of seven chambers. The overall reactor dimensions are 24 cm (long) × 10 cm (width) × 12 cm (depth). Approximately 1 kg of local KSA soil was artificially spiked with kerosene, heavy metals (Cu, Cr, Cd, Pb, Zn, and Hg), and phenol at predetermined concentrations. Thorough mixing was done using mechanical mixer (Gilson Company Inc.) so as to achieve a homogeneous distribution of the contaminants around the soil matrix. The mixed spiked soil was placed in a fume-hood for drying over a period of time necessary to evaporate the solvents (hexane and distilled water). Distilled water was added to adjust the final moisture content of the soil to about 33–70%. The initial conditions of the soil pH, moisture content, organic matter, and electrical conductivity were measured as well as the actual initial concentrations of the contaminants. Then, the uniformly mixed contaminated soil was placed into the cell layer by layer. Each layer was compacted with stainless steel spatula so that the amount of void space was minimized. The reactor used for the experiments consists of the cell, two graphite electrodes serving as anode and cathode, DC power supply (LG, GP-505), processing fluid reservoirs, heavy duty recirculation pump (BVP Instratec), portable data logger (TDS-303, Tokyo Sokki Kenkyujo Co., Ltd) for real-time monitoring of temperature, electric current, and voltage across the system (Figure 2). The two electrode compartments with 240 mL working volume, placed at each end of the cell, were isolated from the soil zone by a porous Perspex plate and filter paper. The conditioning of the electrolyte was controlled using anolyte (2N NaOH) and catholyte (1N HNO$_3$). The pH of the processing fluids was monitored every 8 hr for the 21-day duration of each test. Based on the pH and volume the processing fluids remaining in the electrode chambers, complete replacement, or refill were carried out accordingly. Two planar-shaped electrodes, 10 cm × 10 cm × 0.5 cm, were used to generate a uniform electric field. Within the described cell, two treatment zones that cut across the cell vertically bracketing the spiked soil compartment were filled with the GAC. The data monitoring system was recording electric current variation, applied voltage, and temperature of the soil compartments online following a 30 min preset time step and automatically stores them for subsequent retrieval using floppy disc which can be read using personal computer for easy data and energy consumption analysis. The power supply provides a constant DC electric voltage for the electrokinetic tests. Every week, fractions of the soil specimens were taken at the center of each chamber to determine the residual concentrations of the contaminants, soil pH, water content, organic matter, and electrical conductivity. Upon the completion of each test, the electrode assemblies were disconnected and the soil specimen was extruded from the cell, sectioned into parts, weighed, and preserved in glass vials for organic extraction, heavy metal digestion, and subsequent analyses using the analytical procedures outlined below.

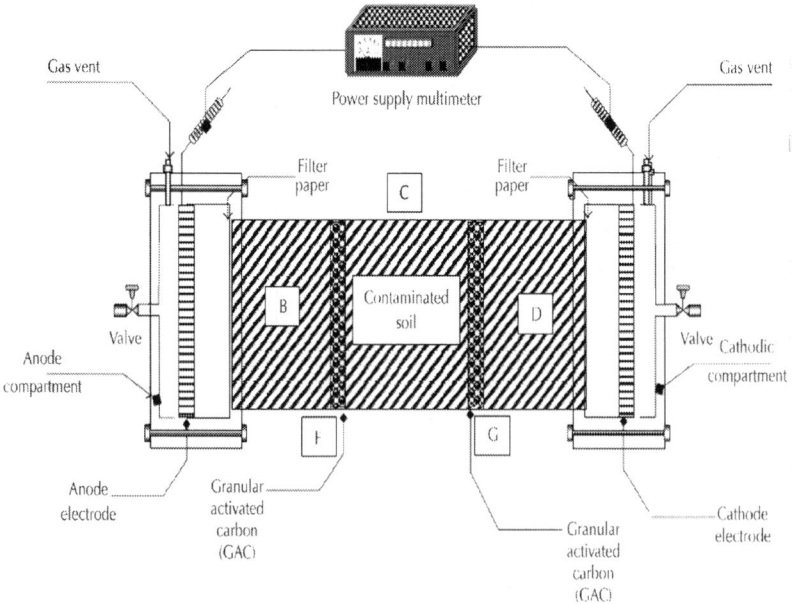

Figure 2: Coupled electrokinetics-adsorption experimental setup.

Analytical Procedures for Contaminant Extraction and Analysis

Heavy Metals. Extraction of heavy metals from soil samples was performed according to guidelines spelt out in EPA method 3050B for acid digestion of soils, sediments, and sludges [48] and analyzed using flame atomic absorption spectrometry (AAnalyst 700, Perkin Elmer). All soil samples were extracted in duplicate. EPA method 7000B [49] was employed for heavy metal analysis using flame atomic absorption spectrometry except for mercury which was analyzed using mercury analyzer (Solid Mercury Analyzer SMS 100, Perkin Elmer) according to EPA method 7473 [50]. Visual MINTEQ 3.0 [51] was employed to model the metal ion speciation using its dissolved concentration, pH, temperature, and ionic strength.

Kerosene and Phenol. A mixture of methylene chloride and hexane (1 : 1) (v/v) was used as the extraction solvent. Soil samples were extracted using pressurized fluid extraction according to EPA method 3545 procedures [52] using accelerated solvent extractor (ASE 200, Dionex). Volume of extract generated was then injected into the GC-MS (Clarus 580, Perkin Elmer) equipped with autosampler for analysis. TPH quantification was done by using the total chromatographic area counts using retention time range for the elution of hydrocarbon within the kerosene range C_8–C_{16}. Guidelines spelt out in EPA mMethod 8270D [53] for the quantification of semivolatile organics by GC-MS were adhered to.

Data Reliability: QC Protocols, Accuracy, and Precision

To evaluate reliability of the analytical procedures, duplicate samples were analyzed for each sample. Quality control (QC) protocols spelt out in EPA method 7000B [49] were used. These include the use of initial calibration blank (ICB), initial calibration verification (ICV), continuous calibration verification (CCV), and continuous calibration blank (CCB). The accuracy of the spiked soil samples was evaluated using percent recovery set at about ±30% of the spiked value [49]. Repeatability of the experimental results was assessed by ensuring that the precision obtained using the relative percent difference (RPD) was not above 30%.

Testing Program and Mathematical Model Development

Box-Behnken design (BBD) was chosen for the experimental design because of its advantages over central composite design (CCD) and 3-level factorial design when dealing with only three factors. In BBD, the experimental points are hyperspherically arranged, equidistant from the central point [38]. Response surface methodology (RSM) was used in modeling, optimization, and interpretation of the results with the help of Design-Expert version 8 (Stat-Ease, Inc.) platform [39, 54]. The investigated variables (called factors in RSM) are the polarity reversal rate, voltage gradient, and initial contaminant concentration designated as A, B, and C, respectively. These variables were selected based on their known influence on contaminant remedial efficiency and were coded and varied according to Table 3. Based on the factor levels and the chosen number of central points (3), a total of fifteen (15) experiments were randomly designed, using the BBD (Table 4), and subsequently conducted. Only one central point is shown in Table 4.

Table 3: Codification and ranges of factors

Variable	Designation	Units	Coded variable levels		
			−1	0	+1
Polarity reversal	A	hr	0	24	48
Voltage gradient	B	V/cm	0.2	0.6	1
Concentration	C	mg/kg	20	60	100

Table 4: Design of experimental runs using the Box-Behnken design

Run order	Polarity reversal,(hr)	Voltage gradient, (V/cm)	Concentration,(mg/ kg)	Remedial efficiency, %
1	0	0.6	20	0.00
2	48	0.6	20	0.00
3	24	1	20	0.00
4	24	1	100	0.00
5	24	0.6	60	79.97
6	0	1	60	72.73
7	24	0.2	20	0.00
8	0	0.2	60	36.93
9	48	1	60	65.66
10	48	0.6	100	0.00
11	0	0.6	100	25.50
12	24	0.2	100	0.00
13	48	0.2	60	34.88

This experimental design was preliminarily evaluated using variance inflation factor (VIF) to check for orthogonality (independence of factors) and leverage which quantitatively measures the potential of a design point to have significant influence on model fit [54]. These were determined using (1) and (2), respectively. VIF value of 1 indicates that the factor is orthogonal to all other factors in the design. In a case whereby factors are highly correlated, then, R^2 value becomes a poor indicator of model's predictive ability and it becomes more and more difficult to unravel how each of the investigated factors affect the response. Experimental points having high leverage values close to 1 should influence the model fit by carrying any error (experimental or measurement) into the model; as such, they should be conducted more carefully [39]:

$$VIF = \frac{1}{\left(1 - R_i^2\right)}$$

(1)

$$\text{Leverage} = \frac{P}{n},$$

$$(2)$$

where R_i^2 is the coefficient of determination; P is the number of model terms; and n is the number of experiments.

Following design evaluation, the responses were fitted to a quadratic model which was fine-tuned by removing any insignificant term. This will maximize R^2 and minimize lack of fit. The general quadratic equation for fitting models in RSM is

$$y = \beta_o + \sum_{i=1}^{k} \beta_i x_i + \sum_{i=1}^{k} \beta_{ii} x_i^2 + \sum_{1 \le i \le j}^{k} \beta_{ij} x_i x_j + \varepsilon,$$

$$(3)$$

where y is the response or dependent variable; k is the number of factors; β_o, β_i, β_{ii}, and β_{ij} are the coefficients to be fitted using regression for constant term, linear, quadratic, and interaction parameters, respectively; and x is the variables.

The developed models were evaluated using the rich diagnostic tools provided in Design-Expert which include normal plot of residuals (to test the assumption of normality of residuals), predicted versus actual plot (to test the assumption of constant variance), Box-Cox plot (to check the need for data transformation), and externally studentized residuals (to check the presence of any outlier in the data). The effects of factors were compared at a particular point in the design space using the perturbation plot. Response surface and contour plots were then generated.

Use of Desirability Function in Numerical Optimization

Desirability function, being one of the mathematical methods for computation of critical values (maximum or minimum) and measuring overall success of optimizing multiple responses using geometric mean, was employed for the optimization of trivalent

Cr remedial efficiency. A search for a combination of factor levels which simultaneously satisfies the goals imposed on factors and responses is first carried out, followed by combining these goals into an overall desirability function that ranges from 0 (outside of the optimization limits) to 1 (at the goal). Combining all responses into overall desirability eliminates favoring one response over another. The aim is not to clinch a desirability value of 1 but to find a good set of conditions that will meet all the set goals for each factor and response [55]. This is achieved using numerical optimization algorithms [52].

RESULTS AND DISCUSSION

Discussion of the monitored results obtained after performing thirteen (13) tests with 3 centre points will be focused on geochemical processes affecting sorption/desorption and migration/ removal mechanisms such as the development of acid/base fronts, migration and reactions, dissolution/precipitation, oxidation/ reduction reactions, complexation, and metallic ion speciation. In addition, presentation of the developed mathematical models and discussion on how the factors affect the respective responses will follow.

Single and Competitive Adsorption of Heavy Metals on Clay

Lukman et al. [24, 45] have discussed the physicochemical characteristics of the saline-sodic soil. Additional discussion will be provided in the subsequent sections. Lukman et al. [45] have found out that the adsorptive capacities of Cu and Zn ions are higher in the multicomponent adsorption scenario than in the single component scenario. The adsorption selectivity sequences obtained using the coefficient of distribution for the single and multicomponent scenarios are Cr > Pb > Cu > Cd > Zn and Cr > Cu > Pb > Cd > Zn, respectively [45]. Yong et al. [40] have identified the general factors that influence selectivity sequence to be ionic size or activity, first hydrolysis constant, soil type, and pH of the system. From the multicomponent desorption study, trivalent Cr ions were tightly held by the soil surface, thus having the least percentage desorption, followed by Cd and Cu ions. Reddy and his coworkers [28, 30, 32] have reported that trivalent Cr ions adsorb highly to soil solids and form cationic species that are insoluble over a wide range of pH. This is in line with the present findings by Lukman et al. [45] which revealed high selectivity for the trivalent Cr during multicomponent adsorption and desorption tests.

Soil PH Distribution, Electrical Conductivity, Bipolar Effects, Electroosmotic Flow, and Current

Soil pH and Electrical Conductivity. The soil pH (8.3) indicates that it contains appreciable soluble salts capable of undergoing alkaline hydrolysis such as sodium carbonate [11]. The hydrolysis of calcite and dolomite may be limited by their low solubility, thus producing a pH of about 8–8.2 in soils. In addition, Na^+ ions do not strongly compete with H^+ ions for exchange sites as do Ca^{2+} ions that are strongly and more tightly held on the soil surface. The inability of the displaced Na^+ ions to inactivate OH^- ions results in increased soil pH, which is usually greater than 8.2. Moreover, for a soil whose pH is greater than 8.2, its exchangeable sodium percentage has to be greater than 15 [11]. Presence of calcite and dolomite coupled with alkaline hydrolysis of sodium carbonate gives high electrical conductivity to the soil (15.24 dS/m).

The saline-sodic nature of the soil necessitates the use of processing fluids (2N NaOH and 1N HNO_3) to continuously neutralize the rapidly generated H^+ and OH^- ions at the anode and cathode, respectively. These fluids were monitored every 8 hours and replaced as they degraded. HNO_3 and NaOH are strong acid and base, respectively, and dissociate completely according to the following reactions:

$$HNO_3 \, (l) \longrightarrow H^+ \, (aq) + NO_3^{\,-} \, (aq) \tag{4}$$

$$NaOH \, (aq) \longrightarrow Na^+ \, (aq) + OH^- \, (aq). \tag{5}$$

Because of the electrochemical decomposition of water, OH^- and H^+ ions are produced at the cathode and anode, respectively, as shown in (6) and (7):

$$4H_2O \, (l) + 4e^- \longrightarrow H_2 \, (g) + 4OH^- \, (aq) \tag{6}$$

$$2H_2O \, (l) \longrightarrow O_2 \, (g) + 4H^+ \, (aq) + 4e^-. \tag{7}$$

The electrochemically generated H^+ and OH^- ions due to water electrolysis at the anode and cathode, respectively, are neutralized to form water molecules (8) because of the OH^- and H^+ ions produced from the dissociation of the catholyte and anolyte, respectively, as shown in (5) and (4):

$$H_+ \, (\text{эd}) + OH_- \, (\text{эd}) \longrightarrow H^5O \, (l)^{\cdot} \tag{8}$$

The oxygen and hydrogen gases generated may be vented out, while some amount may go into the soil and alter the redox chemistry [56]. Na^+ and NO_3^- ions migrate into the soil to the opposite electrodes thereby increasing the electrical conductivity as the treatment process progresses. A sustained and variable electroosmotic flow was observed due to the migration of the Na^+ ions, which could enhance the migration of the double layer complexes toward the cathode, while nitrate ions could be involved in complex formation with the cations [28]. This electroosmotic flow will lead to decreasing volume of the anolyte and increasing volume of the catholyte over time. Hence, refilling the anolyte is necessary if it has not degraded completely. In addition, since the processing fluids are finite in volume and the electrochemical decomposition of water at the electrodes is continuous for the test duration, then, a time will be reached when all the ions in the processing fluids have been exhausted. Consequently, rise and fall in catholyte pH and anolyte pH, respectively, are expected before the complete replacement of the processing fluids. Now, OH⁻ ions generated at the cathode according to (6) migrate into soil toward the anode. In this migration process, soil pH rises (Figure 3) and metal hydroxides are formed which could precipitate and reduce the electrical conductivity (Figure 4) and increase current consumption near the cathode [41]. At the same time, soluble hydroxocomplexes are formed with the cations due to complexing property of the hydroxyl ions [24, 57]. On the other hand, hydrogen ions generated at the anode (7) migrate toward the cathode. This process may lead to soil protonation or desorption of indigenous and spiked heavy metals and hence increase the electrical conductivity (Figure 4) [32]. Given the presence of calcite and dolomite in the soil minerals, the developing acid front may be buffered by the carbonate mineral, thereby hindering any fall in the soil pH (Figure 3). From the forgoing discussion, it is clear that there will be an overall increase in the soil pH and electrical conductivity (Figure 4) as the integrated electrokinetics-adsorption remediation progresses. Results obtained for electrokinetic remediation of high buffering capacity glacial till by Reddy and hiscoworkers [30–32, 58] have corroborated these findings. The transient nature of the acid/base front migration and reactions may be responsible for the lower final values of some pH and EC than the preceding 1st or 2nd week values. In addition, the electroosmotic flow (Figure 5) will undoubtedly vary spatially and temporally as it also depends on the soil zeta potential, processing fluids pH, pore fluid viscosity, and permittivity [7, 59–61].

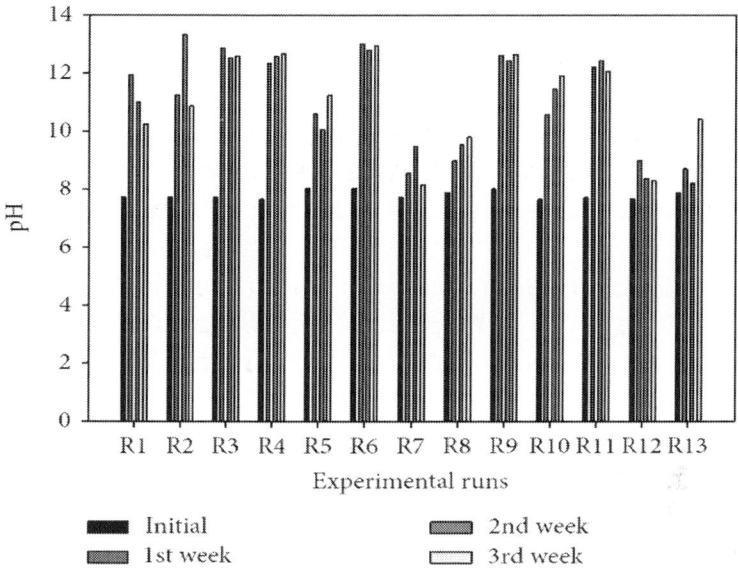

Figure 3: Weekly pH variation.

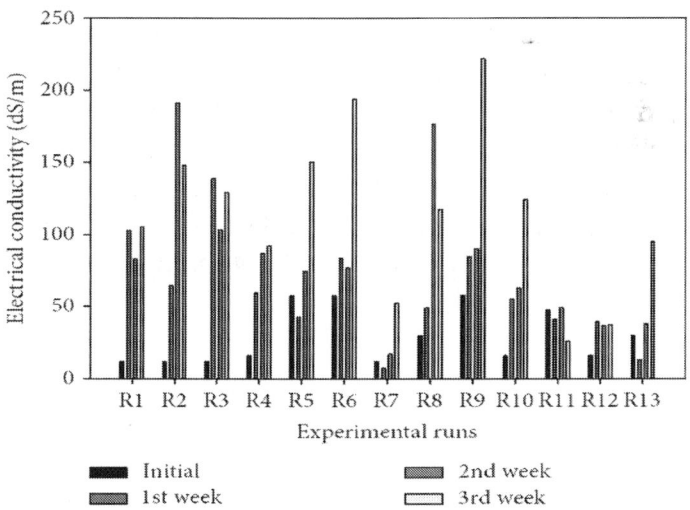

Figure 4: Weekly soil electrical conductivity variation.

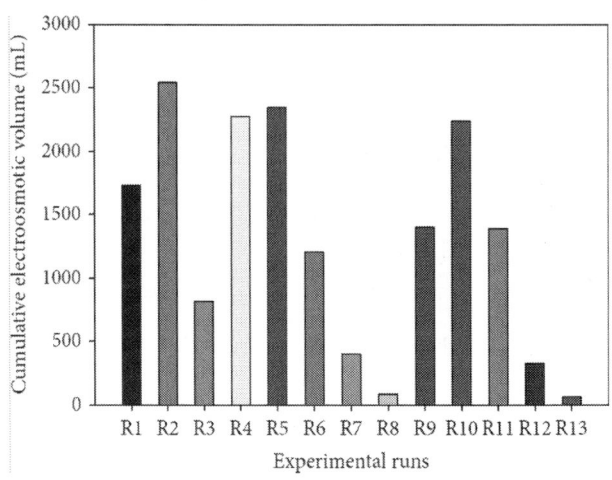

Figure 5: Cumulative electroosmotic volume for each test.

It was observed from Figure 3 that the average initial soil pH after spiking is within the range 7.7–8, lower than the original soil pH (8.3), while the final pH ranges from 8 to 12.9. The lower initial pH was due to the acidity of the contaminant solutions, while the higher final pH values resulted from the high buffering capacity of the soil which neutralized the generated acidic front from (7) but allowed the migration of the basic front generated from (6). In addition, the hydroxyl ions generated from the dissociation of NaOH (5) and reduction of water at the cathode (6) aid in neutralizing the generated acidic front. Consequently, all weekly pH values are higher than the initially spiked soil pH for all the tests. Additionally, low pH rise (8–10.4) was observed for all the tests conducted using 0.2 V/cm (R7-8, R12-13) whereas highest pH (12.6–12.9) was recorded for all tests conducted using 1 V/cm (R3-4, R6, and R9) consistently. High voltage gradient leads to the passage of high amount of current which increases the rate of the electrochemical decomposition of the electrolyte and enhances subsequent migration of the basic front into the soil. This basic front migration is responsible for raising the soil pH. This observed effect of the voltage gradient on the soil pH has been successfully modeled mathematically and the coded linear model equation at 5% significant level (0.05 p value) is presented in (9) while the graphical presentation of the significant influential factors together with 3D response surface and contour plots is given in Figures 6(a) and 6(b):

$$\text{Soil pH} = 11.07 + 0.097 * A + 1.77 * B + 0.39 * C \tag{9}$$

where A is the polarity reversal, hr; B is the voltage gradient, V/cm; and C is the concentration, mg/kg.

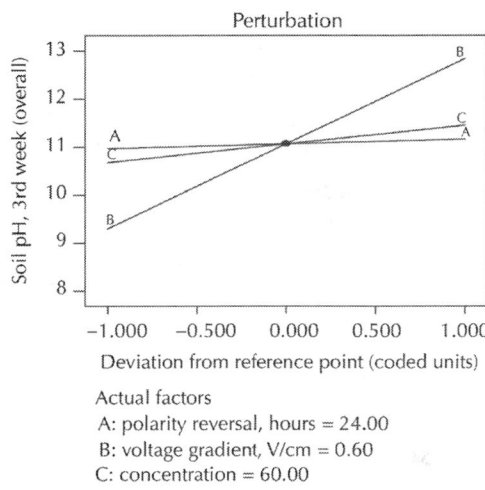

(a)

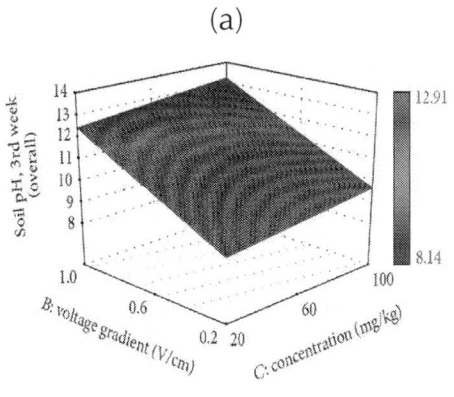

(b)

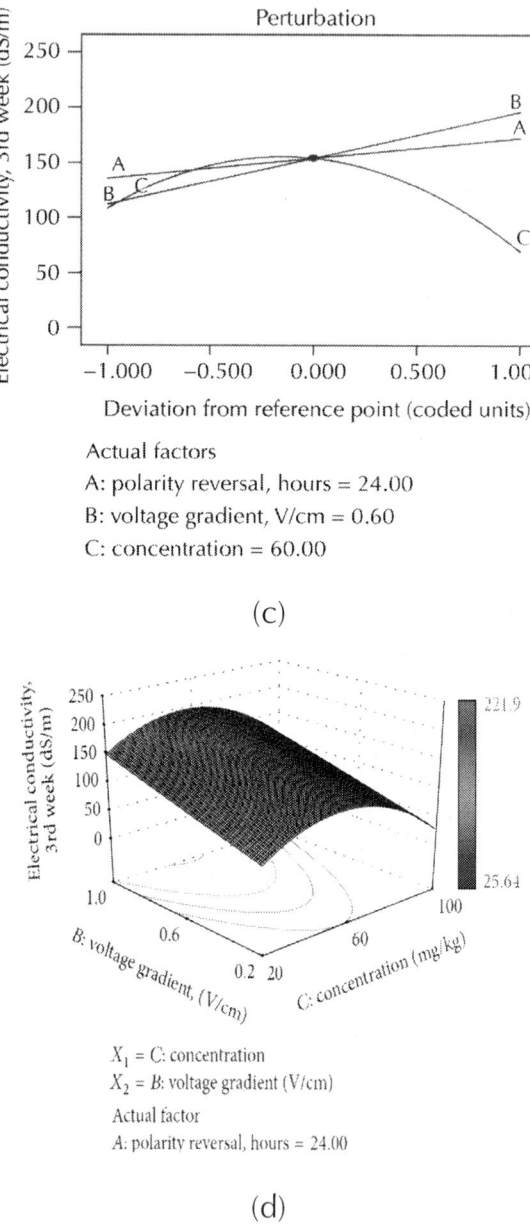

Figure 6: Perturbation plots showing the relative significance of factors on soil pH (a) and electrical conductivity (c) (left). 3D response surface and contour plots showing how the influential factors affect soil pH (b) and electrical conductivity (d) (right).

Anderson and Whitcomb [39] have reported that R^2 is biased; hence, a more accurate, less biased, and better goodness-of-fit statistic called adjusted R^2 was computed for evaluating the model accuracy. The model's R^2 and adjusted R^2 (unbiased estimate of the coefficient of determination) are 0.7725 and 0.7105, respectively. High values of R^2 are essential for modeling the experimental design space, while in identification of significant factors R^2 value does not matter and for significant factors will remain significant [39]. It is very clear that model equation, perturbation, and 3D response surface plots have shown the significant influence of voltage gradient on the soil pH over the other factors (polarity reversal rate and initial contaminant concentration). The relative contribution or effect of any given model term is directly proportional to its coefficient. Perturbation plot (Figure 6(a)) revealed a sequence of relative influence of the operating parameters on the target response as follows: voltage gradient > concentration > polarity reversal.

Bipolar Effects. The two treatment zones F and G contain 100% granular activated carbon which may be used as electrode material due to its electrical conducting properties [14]. The sides of the GAC chambers facing anode and cathode electrodes tend to behave as bipolar electrodes by acting as cathode and anode while the inner sides behave as anode and cathode, respectively. These bipolar electrodes would be expected to generate H^+ and OH^- ions depending on whether the side is acting as anode or cathode [20] and may be expected to alter the pH distribution in the soil profile. These bipolar effects were investigated at the end of R11 and the pH profile is presented in Figure 7. The pH profile shows the variation of pH within the unspiked chambers B and D, spiked chamber C, and GAC chambers F and G. The pH ranges from 11.9 (near the anode) to 12.6 (near the cathode) which suggest that bipolar effects did not manifest due to the presence of carbonate minerals that impact high acid buffering capacity.

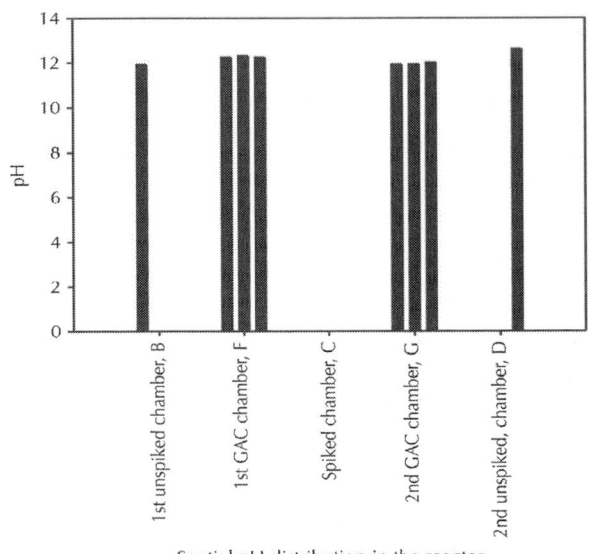

Spatial pH distribution in the reactor

Figure 7: pH profile with two GAC treatment zones for investigating bipolar effects (R11).

Sparks [62] posited that electrical conductivity (EC) is the best index for the assessment of soil salinity. As important as this parameter is, most works on electrokinetic remediation failed to at least report the soil electrical conductivity, let alone monitor its variation over the treatment duration. Electrical conductivity greatly influences electrokinetic remediation, because it determines the amount of current flowing through the soil. The usual voltage gradient of 1 V/cm for bench-scale studies [63] when applied to saline-sodic soils would lead to high electric current flow. Lukman et al. [24] have reported that this would lead to excessive soil heating, reduction in the soil moisture content, high energy and process fluid consumption, high electroosmotic flow rate (Figure 5), and in some cases higher percentage removal of contaminants. EC is simultaneously influenced by many soil properties, viz; water content, soluble salts, grain size, humus, temperature, texture, and cation exchange capacity (CEC) [64]. The 1st week of EC data shows that tests conducted using 1 V/cm (R3, 9, 6) possess the highest EC values with R1 (0.6 V/cm) coming second highest. No discernible trend was visible in the case of initial contaminant concentration despite its influence on the EC as depicted in Figure 6(c). Similar trend was observed for the 3rd week, where R9 and 6 have the highest EC values (Figure 4). A general increase of EC with time and voltage gradient (Figures 6(c) and 6(d)) was observed (except for R11). The reason for this observation has been elaborately discussed above. These variations and impacts of the influential investigated factors have been modeled and presented in the 3D response surface plot in Figure 6(d). Perturbation plot (Figure 6(c)) revealed a sequence of relative influence of the operating parameters on the soil electrical conductivity as follows: concentration > voltage gradient > polarity reversal.

Electroosmotic Flow. The cumulative electroosmotic volume for all the tests presented in Figure 5 shows that R2 (20 mg/kg), R5 (60 mg/kg), and R4 (100 mg/kg) have the highest values. Other parameters that may influence electroosmotic flow are clay zeta potential, voltage gradient, and time-dependent fluid properties such as dielectric constant and viscosity [65]. Equation (10) shows the electroosmotic velocity as derived according to Helmholtz-Smoluchowski (H-S) theory:

$$v_e = \frac{\varepsilon_s \zeta}{\eta} E = k_e E,$$

(10)

where v_e is the electroosmotic velocity; ε_1 is the pore fluid permittivity; η is the pore fluid viscosity; ζ is the soil zeta potential; k_e is the coefficient of electroosmotic conductivity; and E is the voltage gradient.

These parameters make the measured electroosmotic volume for all the tests to vary temporally. The reduction of the thickness of the diffuse double layer resulting from higher ionic concentration with subsequent higher ionic strength causes reduction in the electroosmotic flow [66]; hence higher concentrations usually yield lower electroosmotic volume (Figure 5). Reddy et al. [66] have observed similar trend. The electroosmotic volume usually decreases with time, because of the increase in electrical conductivity with time (Figure 4) that leads to higher ionic strength as the treatment proceeds. Moreover, voltage gradient has been observed to be most influential to the electroosmotic flow (Figure 8). The least electroosmotic volumes recorded belong to the lowest voltage gradient used (0.2 V/cm), that is, in the case of R7, R12, R8, and R13. This is because high voltage gradient causes the passage of high electric current, which leads to high electromigration with subsequent substantial transfer of momentum to the surrounding pore-fluid molecules [66]. The soil zeta potential, defined as the electrical potential existing at the junction between the fixed and mobile parts of the electrical double, is influenced by the type and concentration of dissolved ions in the pore fluid in addition to the pore fluid chemistry. Clay soils, being negatively charged, usually possess negative zeta potential. At low pH below the point of zero charge (PZC), zeta potential may become positive because of excessive protonation and increase in ionic strength resulting from increased dissolution of metal ions in the pore fluid and their subsequent adsorption onto the soil particles and compression of the electrical double layer [67]. Reversal of the zeta potential charge could reverse the direction of the electroosmotic velocity as shown in (10). At high pH values, such as those encountered in this study, deprotonation and metal hydroxide precipitation could maintain a negative zeta potential; hence, electroosmotic flow will remain unidirectional as observed in all the tests. Electroosmotic flow has not been influenced by hydraulic gradient in this study as it occurs even under negative hydraulic. Equation (11) presents the model equation ($R^2 = 0.946$ and adjusted $R^2 = 0.9057$) relating the electroosmotic volume to the factors. Voltage gradient appears to be the most influential, followed by polarity reversal rate and initial contaminant concentration (Figure 8(a)). At high voltage gradient (1 V/cm), the decrease in the electroosmotic volume (Figure 8(b)) may be attributed to the development of bubbles within the electrode chambers, due to temperature rise, which then seeps into the soil to reduce the soil saturation with subsequent reduction in the electroosmotic volume [24]:

$$= 49 + 2.57 * A$$

$$+ 11.68 * B + 1.22 * C$$

$$+ 5.26 * B * C - 5.34 * A^2$$

$$- 20.95 * B^2. \tag{11}$$

Current and Temperature. Table 5 presents the average electric current recorded for each test during the 3-week test duration in descending order of magnitude to show how it is influenced by the applied voltage gradient and how it correspondingly affects the soil pH. Clearly, the higher the voltage gradient, the more amount of current is passed through soil which results in rapid generation of H^+ and OH^- ions and subsequent rise in soil pH (Table 5).

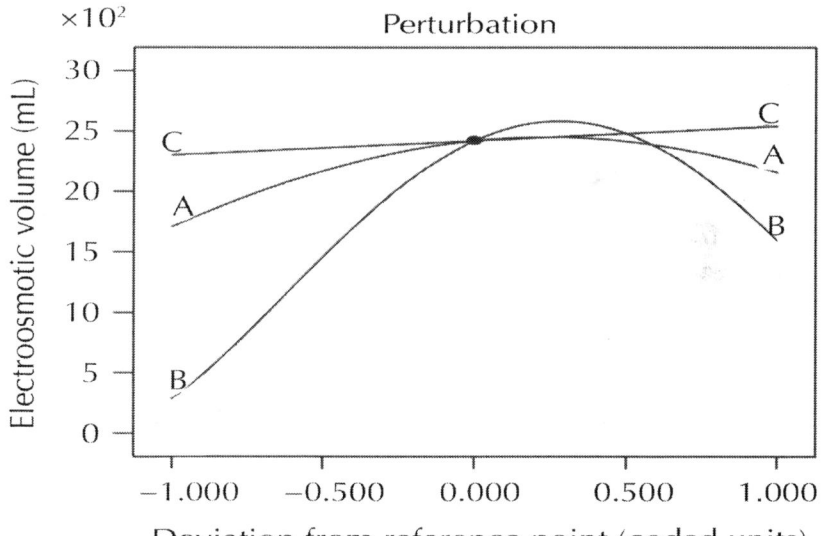

Actual factors
A: polarity reversal, hours = 24.00
B: voltage gradient, V/cm = 0.60
C: concentration = 60.00

(a)

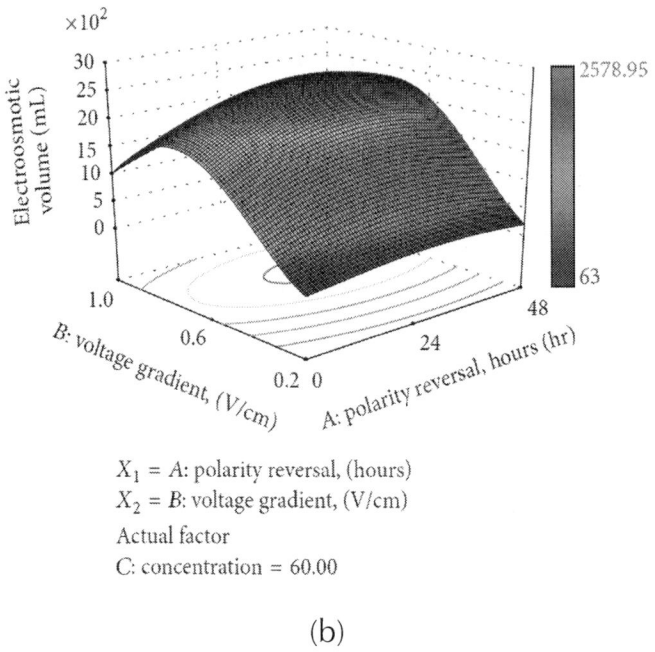

X_1 = A: polarity reversal, (hours)
X_2 = B: voltage gradient, (V/cm)
Actual factor
C: concentration = 60.00

(b)

Figure 8: (a) Perturbation plot showing the relative significance of factors on electroosmotic volume. (b) 3D response surface and contour plots showing the influence of voltage gradient on cumulative electroosmotic volume.

The current is usually low at the beginning of the tests (Figure 9(a)), rises gradually as the tests continue, and then declines, sometimes to a stable value, while in some instances, keeps on fluctuating according to the time-dependent geochemical processes taking place such as ionic dissolution and precipitation and degradation of the processing fluids. Study conducted by Maturi and Reddy [68] corroborated the fluctuating current trend. Upon application of the driving force, the voltage gradient, the processing fluids, and pore fluid migrate while the dissolved ions electromigrate to opposite poles. These processes lead to increase in the ionic strength of the pore fluid thereby increasing the current flow to a maximum value. The observed decline of the current to a stable value may be attributed to the electromigration of cations and anions to the respective electrode with subsequent possible precipitation of the cations due to increase in the soil pH as the test progresses [66, 69]. Temporal geochemical processes such as mineral and chemical dissolution and neutralization reactions taking place in the electrode chambers also contribute to the variation of the electric current. A maximum value of 5.13 A was recorded for R6 whose average current was 3.02 A. This current is considered extremely high, considering the fact that it is about two orders of magnitude greater than the recorded current values for other bench-scale studies that employed the Lasagna process (<30 mA) in other soil apart from saline-sodic soil as shown in Table 1. Other studies using electrokinetic remediation only using voltage gradient of 1 V/cm or higher have reported higher values but usually less than 300 mA [58, 66, 70]. Using low voltage gradient of 0.2 V/cm has only resulted in reducing the current to about 130–210 mA (Table 5). This unique and important observation may be explained by the high salinity and sodicity of the investigated soil which provides large amount of dissolved salts and minerals (carbonates) in the pore fluid for sustained high electrical conduction. High current flow through the soil will significantly affect the soil temperature, electroosmotic flow rate, electrode material and processing fluids degradation, soil pH, geochemical processes, remedial efficiency and energy consumption. In a related study by Lukman et al. [24], they also recorded similar high current (2.8 A). To emphasize on the effect of the electric current on the soil temperature, current and temperature readings recorded using a time step of 30 min is presented in Figure 9 for R11 (voltage gradient = 0.6 V/cm). This test has 0.61 A and 28.45°C as the average current and temperature respectively. The maximum values were 0.91 A and 34.6°C respectively which were recorded under room temperature of 24°C. It is clear from Figure 9 that low current leads to low soil temperature and vice-versa. In a preliminary study conducted by Lukman et al. [24] using 1 V/cm, 36.34°C, and 47°C were the average and maximum soil temperatures, indicating that the soil becomes very hot when using 1 V/cm. While soil heating may be advantageous in increasing the volatility of organics, solubility of minerals (carbonates), and reduction in pore fluid viscosity which will increase electroosmotic flow, it may also be undesirable since it will reduce the soil moisture content due to pore fluid evaporation with subsequent reduction in current and electroosmotic flow. In addition, it will increase soil electrical conductivity and energy expenditure [20]. Previous studies have not reported significant rise of soil temperature during bench-scale tests [20]. A linear model was obtained (12) which relates the factors to the average electric current whose respective R^2 and adjusted R^2 are 0.9556 and 0.9435. The perturbation and response surface plots (Figure 10) also revealed the significant influence of the applied voltage gradient over initial contaminant concentration and polarity reversal rate:

Sqrt (Average current)

$$= 1 + 0.020 * A + 0.59 * B - 0.059 * \qquad (12)$$

Table 5: Comparing electrical current with voltage gradient and soil pH for all tests

Run	Current, A	Voltage gradient, V/cm	pH
R6	3.02	1	12.9
R9	2.65	1	12.6
R3	2.25	1	12.6
R4	2.04	1	12.7
R2	1.32	0.6	10.9
R1	1.17	0.6	10.2
R10	1.12	0.6	11.9
R5	1.03	0.6	11.2
R11	0.61	0.6	12.0
R8	0.21	0.2	9.8
R12	0.15	0.2	8.3
R7	0.14	0.2	8.1
R13	0.13	0.2	10.4

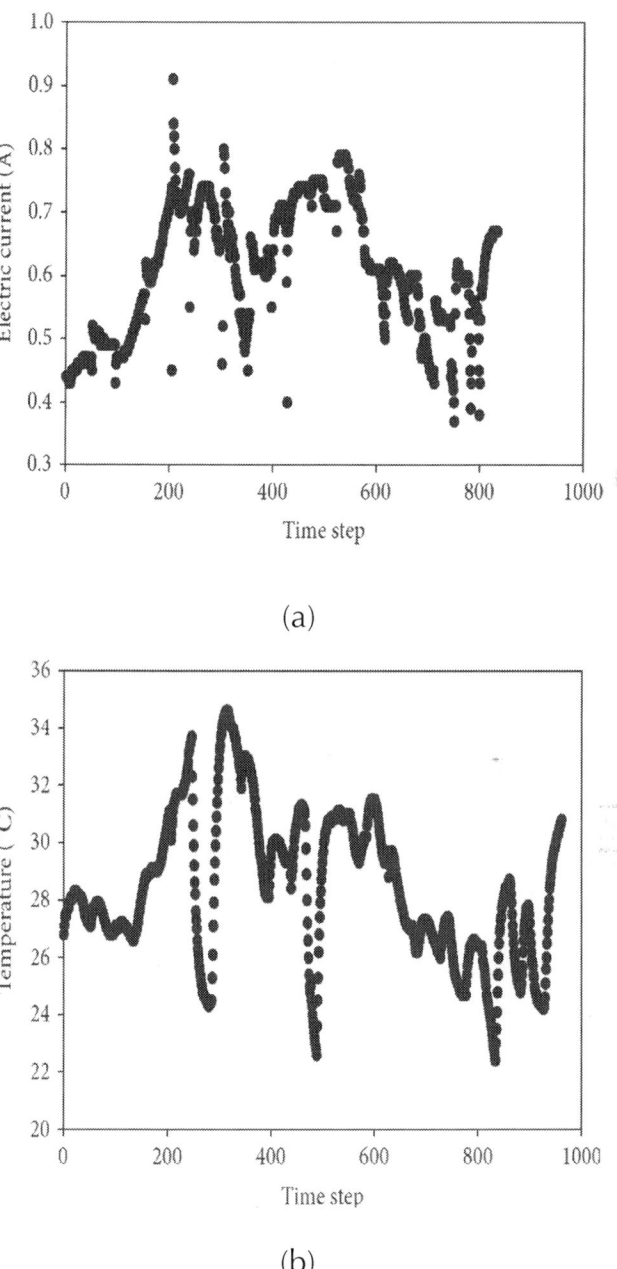

(a)

(b)

Figure 9: Comparing variations of electric current with soil temperature: (a) current; (b) temperature.

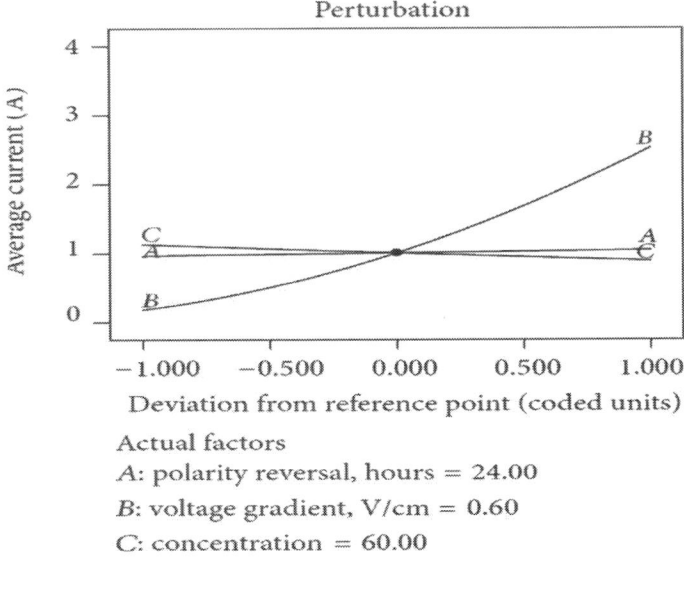

Perturbation

Actual factors
A: polarity reversal, hours = 24.00
B: voltage gradient, V/cm = 0.60
C: concentration = 60.00

(a)

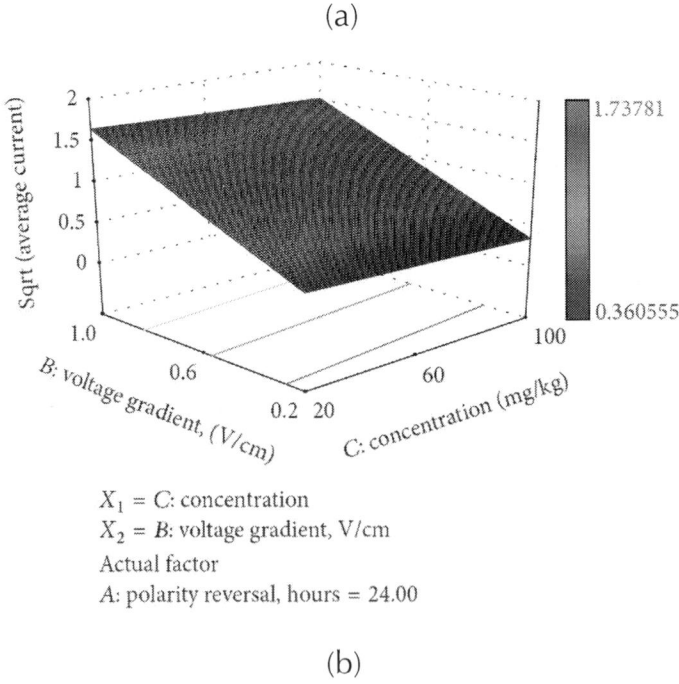

X_1 = C: concentration
X_2 = B: voltage gradient, V/cm
Actual factor
A: polarity reversal, hours = 24.00

(b)

Figure 10: (a) Perturbation plot showing the relative significance of factors on average electric current. (b) 3D response surface and contour plots showing the influence of voltage gradient on average electric current.

Trivalent Chromium Migration, Model Validation, and Optimization

Figure 11 presented the distribution and migration of trivalent Cr from the contaminated chamber, C, to the GAC chambers F and G for all the thirteen (13) tests. This migration becomes more pronounced for tests R5, R6, and R9. In the case of R6 (no polarity reversal), significant trivalent Cr migration took place from the contaminated chamber, C, to the GAC chamber, F, near the anode. This observation may be attributed to the formation of high amount of negatively charged metal hydroxocomplexes at pH 12.9, which are then attracted to the anode via electromigration but become adsorbed onto the GAC in chamber F during the transport process. Visual MINTEQ 3.0 [51] was employed to model the trivalent Cr ion speciation for R5 from the weekly monitoring data using the dissolved concentration, pH, temperature and ionic strength. The speciation diagram presented in Figure 12 reveals the increasing dominance of the negatively charged complex and the decreasing concentration of aqueous $Cr(OH)_3$ at pH 11.2. This explains the greater movement of the trivalent Cr species toward the anode in R6 at pH 12.9. Pourbaix [71] and Chinthamreddy and Reddy [29] have already asserted that $Cr(OH)_4^-$ ions will become the dominant species at pH values greater than 11.8, thus, trivalent Cr solubility increases. However, under normal soil pH, trivalent Cr has limited solubility and highly adsorbs to soil [29, 32]. In a related study by Reddy and Chinthamreddy [30] which involved an alkaline and high acid buffering soil called glacial till, they did not observe significant trivalent Cr migration and no removal. Although, the soil redox state may be dynamic because of the generation of oxygen and hydrogen gases at the electrodes in addition to the possible presence of iron (reducing agent), manganese (oxidizing agent) or microorganisms that can oxidize the trivalent Cr to the hexavalent form; oxidation of trivalent Cr does not take place appreciably in high buffering capacity soil such as saline-sodic soil [28]. For this reason, hexavalent Cr was not studied. Migration of the trivalent Cr from the contaminated chamber to the GAC chambers indicated remarkable remedial efficiency for some of the tests (R5, R6 and R9) while others indicated low or no removal at all (R1–R4, R7, R10, and R12). There is zero remedial efficiency when there was accumulation of the contaminant at the sampling location thereby having the residual concentration (C_o) to be greater than the initial (C), in which case, $C_o/C > 1$. Hence Figure 11 utilized C_o/C to indicate the migration of trivalent Cr when $C_o/C < 1$ or its accumulation at any given location or chamber when $C_o/C > 1$.

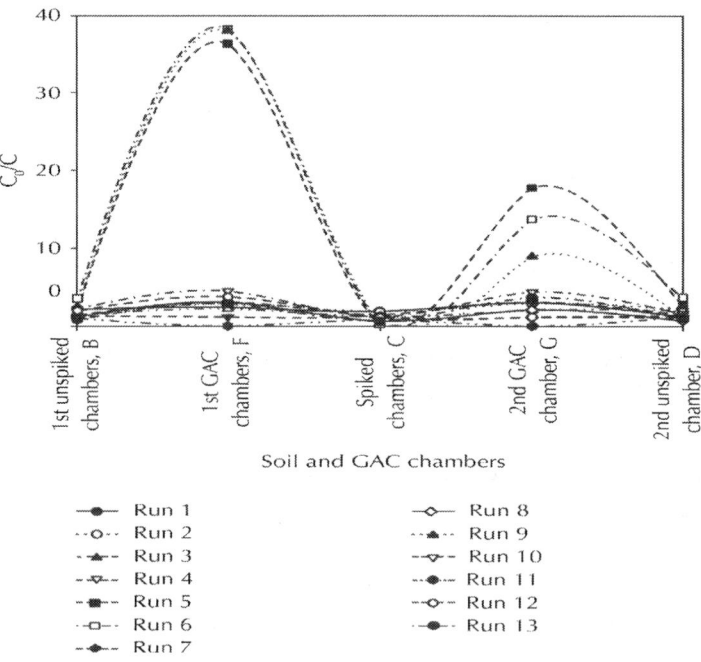

Figure 11: Trivalent Cr distribution and migration from the contaminated chamber to the GAC chambers after 13 tests.

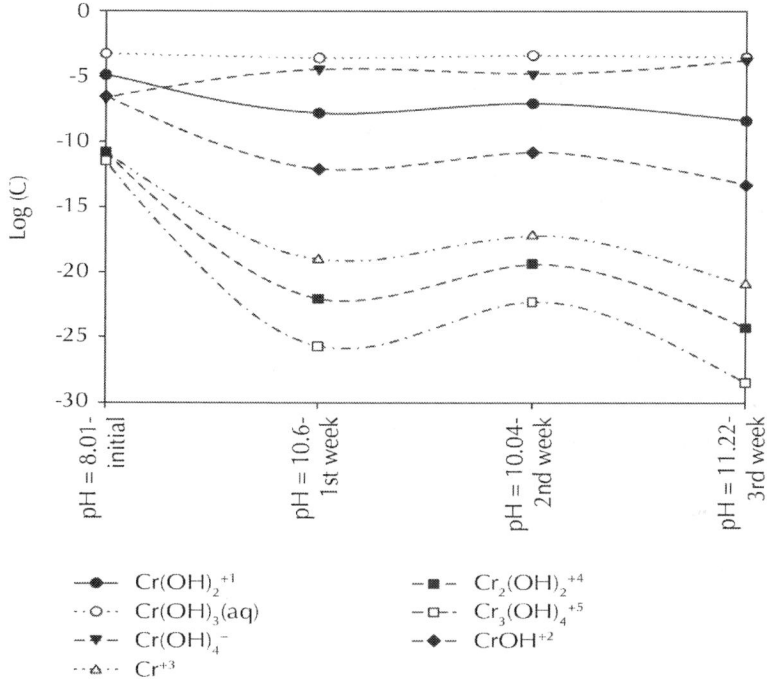

Figure 12: Speciation diagram for trivalent Cr species at different weekly pH values.

Mass balance analyses of Cr were performed for Runs 8, 11, and 13. From Table 6, the mass balance for Runs 8, 11, and 13 is 121.75, 74.51, and 148.99%, respectively. These values were obtained using the ratio between the residual Cr in the contaminated chamber (C) plus any increase in Cr concentration in the GAC chamber and the initial Cr concentration. Among other reasons for the discrepancies in mass balance that is sometimes encountered during electrokinetic remediation as put forward by previous investigators [30, 66, 72] include adsorption onto the electrode and geotextile materials (which houses the GAC in the two chambers) and non-uniform distribution of contaminants within the small soil sample (about 2 g) taken for acid digestion and analysis. Taking different samples spatially from the contaminated chamber will help improve the mass balance.

Table 6: A sample mass balance analysis of trivalent Cr for Runs 8, 11, and 13

Runs	Run 8	Run 11	Run 13
Initial concentration, mg/kg	37.20	77.95	37.20
Residual concentration, mg/kg	23.46	58.08	24.23
1st GAC chamber, F			
Initial concentration, mg/kg	6.90	6.90	6.90
Residual concentration, mg/kg	21.15	0.00	19.70
2nd GAC chamber, G			
Initial concentration, mg/kg	6.90	6.90	6.90
Residual concentration, mg/kg	14.48	0.00	25.30
Mass balance, %	121.75	74.51	148.99

The tests were sorted in decreasing order of remedial efficiency (Table 7) to reveal some salient points that will help in providing adequate connection between factors and responses. Highest remedial efficiencies (79.97–34.88%) were recorded for tests involving 60 mg/kg initial trivalent Cr concentration, whereas no removal was recorded for all tests involving 20 mg/kg. Only one test involving 100 mg/kg recorded some remedial efficiency (Table 7). Low remedial efficiency at 20 mg/kg may be attributed to the availability of adsorption sites for trivalent Cr ions coupled with the high selectivity for Cr for this particular soil type [45] at the given concentration. At higher concentrations (100 mg/kg) and pH, trivalent Cr may precipitate as $Cr(OH)_3$, thus, rendering it immobile [66]. Even with low electric current, electroosmotic flow and voltage gradient (0.2 V/cm), 34.88% and 36.93% of the trivalent Cr was removed from the contaminated chamber in tests R13 and R8, respectively. Polarity reversal rate did not show any discernible pattern. Hence, there is need for simultaneous optimization of these three factors for optimal removal of the trivalent Cr. It is important to note that high voltage gradient (1 V/cm) or passage of high electric current does not necessarily translate into high remedial efficiency but will definitely increase the energy expenditure. At high voltage gradient, current is high, leading to high electroosmotic flow toward cathode. This opposite flow may interfere with the electromigration of the anionic trivalent Cr species that are migrating toward the anode, thus, reducing the overall remedial efficiency. Electromigration constitute the major transport mechanism for charged species whose rate is 10–300 times higher than the advective electroosmotic transport [73]. At low voltage gradient (0.2 V/cm), extremely low electroosmotic flow takes place and sustained electromigration prevails. The weekly percentage removal of trivalent Cr from the contaminated chamber is presented in Figure 13. The dynamic and temporal changes in the geochemical processes controlling the contaminant removal are attributable to the observed trends in the weekly percentage removal.

Table 7: Comparing trivalent Cr remedial efficiency with factors and some responses

Runs	Remedial	Cur-rent,	Residual,	Electro-osmotic volume,	Polarity reversal	Volt-age gradi-ent,	Initial Cr
	efficien-cy, %	A	pH	mL	rate, hr	V/cm	concen-tration, mg/kg
R5	79.97	1.03	11.2	2344.50	24	0.6	60
R6	72.73	3.02	12.9	1201.50	0	1	60
R9	65.66	2.65	12.6	1399.50	48	1	60
R8	36.93	0.21	9.8	81.00	0	0.2	60
R13	34.88	0.13	10.4	63.00	48	0.2	60
R11	25.50	0.61	12.0	1387.84	0	0.6	100
R1	0.00	1.17	10.2	1728.00	0	0.6	20
R2	0.00	1.32	10.9	2542.50	48	0.6	20
R3	0.00	2.25	12.6	814.50	24	1	20
R4	0.00	2.04	12.7	2272.50	24	1	100
R7	0.00	0.14	8.1	396.00	24	0.2	20

| R10 | 0.00 | 1.12 | 11.9 | 2236.50 | 48 | 0.6 | 100 |
| R12 | 0.00 | 0.15 | 8.3 | 324.00 | 24 | 0.2 | 100 |

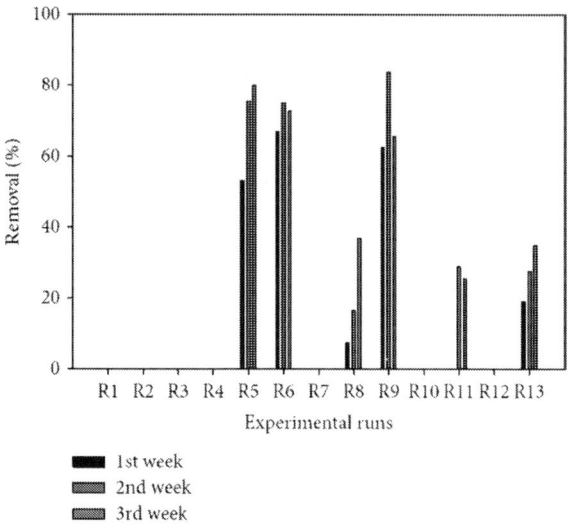

Figure 13: Weekly percentage removal of trivalent Cr for 13 tests.

Equation (13) relates the investigated factors to the remedial efficiency with 0.9335 and 0.8966 as the R^2 and adjusted R^2 values, respectively:

$$\text{Sqrt (Cr, remedial efficiency)}$$

$$= 8.78 - 0.71 * A + 0.58 * B$$

$$+ 0.63 * C - 1.50 * B^2 - 7.39 * C^2. \tag{13}$$

Perturbation plot (Figure 14(a)) also supports the observed influence of the initial Cr concentration on the remedial efficiency, followed by voltage gradient, then, polarity reversal rate. The investigated factor levels can be used to determine the optimal conditions required to achieve maximum remedial efficiency as depicted in the 3D response surface plot (Figure 14(b)).

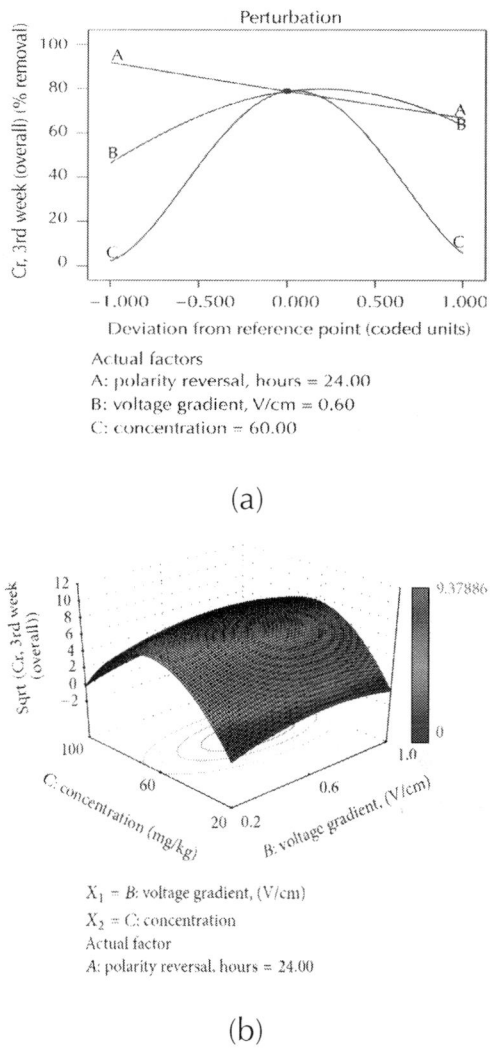

(a)

(b)

Figure 14: (a) Perturbation plot showing the relative significance of factors on trivalent Cr remedial efficiency. (b) 3D response surface and contour plots showing the influence of initial contaminant concentration on trivalent Cr remedial efficiency.

Model Validation. To validate the practical applicability of the developed models affecting the remedial efficiency (13) and soil pH (9), additional experimental test was run at voltage gradient of 1 V/cm, initial contaminant concentration of 44.15 mg/kg, and without polarity reversal (Table 8). Results of the model validation showed that the experimental results lie within 90% confidence interval (CI) and prediction interval (PI) with associated prediction error of 2.35% and 32.64% for soil pH and remedial efficiency, respectively. Since the validation results fall within the prediction interval, then, the outcome of the confirmation test was a success [39]. Hence, the models can provide good approximations necessary to move in the proper direction.

Table 8: Experimental validation of trivalent Cr remedial efficiency and soil pH using voltage gradient = 1 V/cm; average concentration = 44.15 mg/kg; and polarity reversal rate = 0 hr

Response	Experi-mental result	Model predic-tion	Predic-tion error, %	90% CI* low	90% CI high	90% PI** low	90% PI high
Cr, re-medial efficiency	75.88	51.11	32.64	31.17	75.95	18.36	100.00
Residual soil, pH	12.3	12.6	2.35	11.7	13.5	10.8	14.0

Optimization of Trivalent Chromium Removal. Numerical optimization was employed to find the optimal factor levels that will specifically target maximum remedial efficiency of trivalent Cr while optimizing all the other contaminant remedial efficiencies and responses (Figure 15). An overall desirability value of 0.715 was obtained and its variation based on the influential factors (initial concentration and voltage gradient) is depicted in Figure 16. Optimal conditions required to achieve effective trivalent Cr removal at 60 mg/kg are presented in Table 9. Overall desirability of 0.715 was attained at the following optimal conditions: voltage gradient = 0.36 V/cm; polarity reversal rate = 17.63 hr; soil pH = 10.0. Under these conditions, the expected trivalent Cr remedial efficiency is 64.75%.

Table 9: Optimal factor levels required to maximize remedial efficiency of trivalent Cr

Item	Value
Polarity reversal, hours	17.63
Voltage gradient, V/cm	0.36
Concentration, mg/kg	60.00
Expected remedial efficiency of trivalent Cr	64.75
Expected residual soil pH	10.00
Desirability	0.715

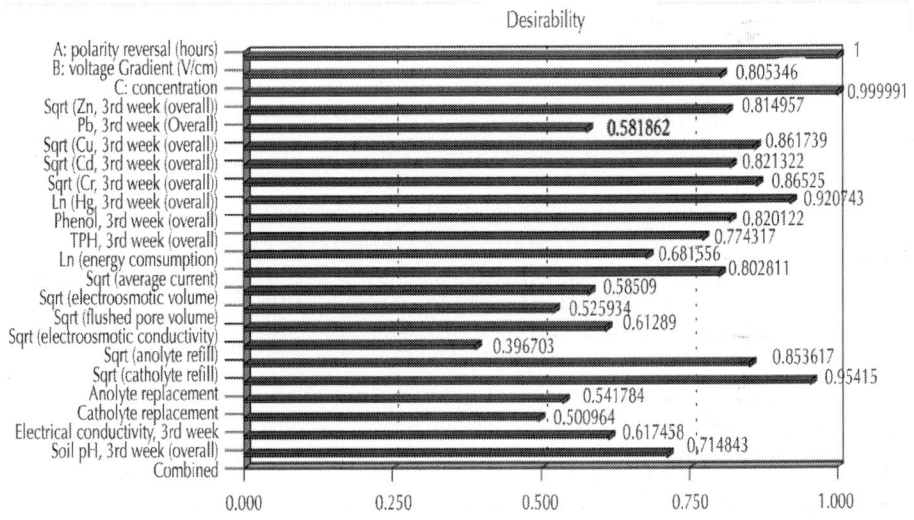

Figure 15: Combined and individual response desirability values for all responses and factors.

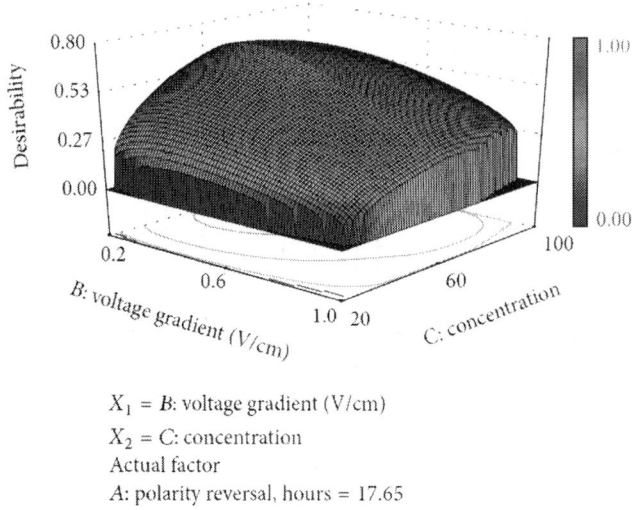

$X_1 = B$: voltage gradient (V/cm)
$X_2 = C$: concentration
Actual factor
A: polarity reversal, hours = 17.65

Figure 16: 3D surface plot of the overall desirability variation relative to influential factors.

Impacts of the Integrated Electrokinetic Remediation on Soil Physicochemical Properties

Preceding sections have elaborately discussed and modeled the impacts of the proposed remediation technique on the soil pH and electrical conductivity. Additionally, the passage of electric current and soil pH gradients will result in the following physicochemical interactions: (1) possible dissolution of the clay minerals beyond a pH range of 7–9; (2) dissolution of available soil salts such as carbonates; (3) production of cementitious products resulting from the precipitation of metal ions at pH values corresponding to their hydroxide solubility values; and (4) soil structural changes which affect its engineering characteristics [41–44]. Surface area, pore volume and size (Table 10), mineralogical compositions (Table 11), and elemental constituents (Table 12) were analyzed, before and after the test for R5. At the end of the test (pH = 11.2), the soil specific surface area has increased (9.07 to 11.21 m²/g) with corresponding increase in the pore volume and size. These results have confirmed that some dissolution of the soil minerals has taken place during the electrokinetic remediation process due to variations in the pore fluid chemistry. Soil pores are due to the presence of interlayer spaces that becomes prominent in 2 : 1 clay mineral types such as montmorillonite and smectite [40,62, 74]. Table 11 presents the mineral transformation where dolomite completely disappeared; calcite and quartz were altered and degraded, respectively, after the test. The constituent soil elements were not spared as the amount of each one either increased or decreased after the test as shown in Table 12. These observations may be explained by microbially-driven biotransformation processes involving dissolution and precipitation, which take place under both aerobic anaerobic conditions. This leads to mineral dissolution and formation of new minerals from aqueous ions (biomineralization) as noticed in Table 11 [40]. Yong et al. [40] have asserted that the scientific basis for biomineralization is still not well understood.

Table 10: Values of soil surface area and pore volume and size, before and after treatment

Description	BET* surface area, m²/g	Pore volume, cm³/g	Pore size, A
Before	9.07	0.014	62.55
After	11.21	0.045	163.24

Table 11: Soil mineralogical transformations before and after treatment

Phase name	Before, %	After, %
Quartz, SiO_2	87.4	55.3
Calcite, $CaCO_3$	5.2	44.7
Dolomite, $CaMg(CO_3)_2$	7.4	—

Table 12: Values of constituent soil elements, before and after treatment

Element	Before, %	After, %
Ca	37.64	42.06
Si	34.73	23.42
Fe	10.41	15.06
Al	7.6	9.55
K	3.42	4.61
Mg	2.48	2.49
Pd	2.85	1.46
Ti	0.86	1.35

CONCLUSIONS

The study reported herein investigated the migration of trivalent Cr ions from a multiple contaminated natural saline-sodic soil. The soil salinity and sodicity, which provided large amount of dissolved salts and minerals (carbonates) in the pore fluid for sustained high electrical conduction, were responsible for the extremely high electric current flow.

This led to excessive soil heating, high energy and process fluid consumption, high electroosmotic volume, and in some cases higher percentage removal of trivalent Cr. Significant migration of Cr from the contaminated chamber to the granular activated carbon chamber was recorded which led to highest remedial efficiencies (79.97–34.88%) for tests involving 60 mg/kg initial trivalent Cr concentration, whereas no removal was recorded for all tests involving 20 mg/kg. Even under low electric current, electroosmotic flow, and voltage gradient (0.2 V/cm), up to 36.93% of the trivalent Cr was removed from the contaminated chamber. It has been shown that high voltage gradient (1 V/cm) or passage of high electric current does not necessarily translate into high remedial efficiency. Bipolar effects did not manifest due to the presence of carbonate minerals that impact high acid buffering capacity. For test without polarity reversal, trivalent Cr moved toward the anode due to the formation of high amount of anionic hydroxocomplex at high pH, which was further attracted to the anode via electromigration. Nonadsorption of this ion onto the negatively charged clay soil due to the possession of similar charge increased its availability and mobility. Speciation modeling using Visual MINTEQ 3.0 reveals the increasing dominance of the anionic $Cr(OH)_4^-$ and the decreasing concentration of aqueous $Cr(OH)_3$ at pH 11.2. Effects of voltage gradient, initial contaminant concentration, and polarity reversal rate on the effective removal of Cr ions were experimentally studied using the Box-Behnken Design of experiment and mathematically modeled and numerically optimized using response surface methodology. Results of the model validation showed that the experimental results lie within 90% confidence interval and prediction interval with associated prediction error of 2.35% and 32.64% for soil pH and trivalent Cr remedial efficiency, respectively. Overall desirability of 0.715 was attained at the following optimal conditions: voltage gradient = 0.36 V/cm; polarity reversal rate = 17.63 hr; and soil pH = 10.0. Under these conditions, the expected trivalent Cr remedial efficiency is 64.75%. Passage of electric current and variations in the pore fluid chemistry led to soil mineral dissolution and alteration via biotransformation.

ACKNOWLEDGMENTS

The authors would like to acknowledge the support provided by King Abdul-Aziz City for Science and Technology (KACST) through the Science & Technology Unit at King Fahd University of Petroleum & Minerals (KFUPM) for funding this work through Project no. 11-Env1669-04, as part of the National Science, Technology and Innovation Plan.

REFERENCES

1. S. V. Ho, B. M. Hughes, P. H. Brodsky, J. S. Merz, and L. P. Egley, "Advancing the use of an innovative cleanup technology: case study of Lasagna," Remediation Journal, vol. 9, pp. 103–116, 1999.

2. P. H. Brodsky and S. V. Ho, "In situ remediation of contaminated soils," U.S. Patent 5.398.756, 1995.

3. S. V. Ho and P. H. Brodsky, "In-situ remediation of contaminated heterogeneous soils," U.S. Patent 5,476,992, 1995.

4. F. F. Reuss, "Sur un nouvel effet de l'électricité galvanique," Mémoires de la Societé Impériale des Naturalistes de Moscou, vol. 2, pp. 327–337, 1809.

5. H. A. Abramson, Electrokinetic Phenomena and Their Application to Biology and Medicine, ACS Monograph Series, Chemical Catalog, New York, NY, USA, 1934.

6. I. Ravina and D. Zaslavsky, "Non-linear electrokinetic phenomena—I: review of literature," Soil Science, vol. 106, pp. 60–66, 1968.

7. A. T. Yeung, "Geochemical processes affecting electrochemical remediation," in Electrochemical Remediation Technologies for Polluted Soils, Sediments and Groundwater, pp. 65–94, John Wiley & Sons, 2009.

8. J. Virkutyte, M. Sillanpää, and P. Latostenmaa, "Electrokinetic soil remediation—critical overview,"Science of the Total Environment, vol. 289, no. 1–3, pp. 97–121, 2002.

9. S.-S. Kim, S.-J. Han, and Y.-S. Cho, "Electrokinetic remediation strategy considering ground strate: a review," Geosciences Journal, vol. 6, pp. 57–75, 2002.

10. V. P. Evangelou, Environmental Soil and Water Chemistry: Principles and Applications, Wiley-Interscience, New York, NY, USA, 1998.

11. I. P. Abrol, J. S. P. Yadav, and F. I. Massoud, "Salt-affected soils and their management, food and agriculture Organization of

the United Nations," FAO Soils Bulletin, vol. 39, 1988.

12. C. D. Palmer and P. R. Wittbrodt, "Processes affecting the remediation of chromium-contaminated sites," Environmental Health Perspectives, vol. 92, pp. 25–40, 1991.

13. A. N. Puri and B. Anand, "Reclamation of alkali soils by electrodialysis," Soil Science, vol. 42, no. 1, pp. 23–27, 1936.

14. S. V. Ho, P. W. Sheridan, C. J. Athmer et al., "Integrated in situ soil remediation technology: the Lasagna process," Environmental Science and Technology, vol. 29, no. 10, pp. 2528–2534, 1995.

15. C. J. Athmer and S. V. Ho, "Field studies: organic-contaminated soil remediation with lasagna technology," in Electrochemical Remediation Technologies for Polluted Soils, Sediments and Groundwater, K. R. Reddy and C. Cameselle, Eds., pp. 625–646, John Wiley & Sons, 2009.

16. K. Maturi and K. R. Reddy, "Simultaneous removal of organic compounds and heavy metals from soils by electrokinetic remediation with a modified cyclodextrin," Chemosphere, vol. 63, no. 6, pp. 1022–1031, 2006.

17. K. R. Reddy, P. R. Ala, S. Sharma, and S. N. Kumar, "Enhanced electrokinetic remediation of contaminated manufactured gas plant soil," Engineering Geology, vol. 85, no. 1-2, pp. 132–146, 2006.

18. T. Li, S. Yuan, J. Wan et al., "Pilot-scale electrokinetic movement of HCB and Zn in real contaminated sediments enhanced with hydroxypropyl- -cyclodextrin," Chemosphere, vol. 76, no. 9, pp. 1226–1232, 2009.

19. M. Elektorowicz, "Electrokinetic remediation of mixed metals and organic contaminants," inElectrochemical Remediation Technologies for Polluted Soils, Sediments and Groundwater, pp. 315–331, John Wiley & Sons, 2009.

20. S. V. Ho, C. J. Athmer, P. W. Sheridan, and A. P. Shapiro, "Scale-up aspects of the Lasagna process for in situ soil decontamination," Journal of Hazardous Materials, vol. 55, no. 1–3, pp. 39–60, 1997.

21. J. W. Ma, H. Wang, and Q. Luo, "Movement-adsorption and its mechanism of Cd in soil under combining effects of electrokinetics and a new type of bamboo charcoal," Chinese Journal of Environmental Science, vol. 28, no. 8, pp. 1829–1834, 2007 (Chinese).

22. J. W. Ma, H. Wang, and R. R. Li, "Removal of cadmium in kaolin by electrokinetics-bamboo charcoal adsorption," Environmental Chemistry, vol. 26, pp. 634–637, 2007 (Chinese).

23. J. W. Ma, F. Y. Wang, Z. H. Huang, and H. Wang, "Simultaneous removal of 2,4-dichlorophenol and Cd from soils by electrokinetic remediation combined with activated bamboo charcoal," Journal of Hazardous Materials, vol. 176, no. 1–3, pp. 715–720, 2010.

24. S. Lukman, M. H. Essa, N. D. Mu'azu, and A. Bukhari, "Coupled electrokinetics-adsorption technique for simultaneous removal of heavy metals and organics from saline-sodic soil," The Scientific World Journal, vol. 2013, Article ID 346910, 9 pages, 2013.

25. S. V. Ho, C. Athmer, P. W. Sheridan et al., "The lasagna technology for in situ soil remediation. 1. Small field test," Environmental Science and Technology, vol. 33, no. 7, pp. 1086–1091, 1999.

26. S. V. Ho, C. Athmer, P. W. Sheridan et al., "The lasagna technology for in situ soil remediation. 2. Large field test," Environmental Science & Technology, vol. 33, no. 7, pp. 1092–1099, 1999.

27. B. D. Swift and J. J. Tarantino, "Application of the Lasagna soil remediation technology at the DOE paducah gaseous diffusion plant," in Proceedings of the Waste Management Symposium (WM ‹03), Tucson, Ariz, USA, February 2003.

28. S. Chinthamreddy and K. R. Reddy, "Oxidation and mobility of trivalent chromium in manganese-enriched clays during electrokinetic remediation," Soil and Sediment Contamination, vol. 8, no. 2, pp. 197–216, 1999.

29. S. Chinthamreddy and K. R. Reddy, "Geochemistry of chromium during electrokinetic remediation," inProceedings of the 4th International Symposium on Environmental Geotechnology and Global Sustainable Development, Boston, Mass, USA, 1998.

30. K. K. Reddy and S. Chinthamreddy, "Effects of initial form of chromium on electrokinetic remediation in clays," Advances in Environmental Research, vol. 7, no. 2, pp. 353–365, 2003.

31. K. R. Reddy and U. S. Parupudi, "Removal of chromium, nickel and cadmium from clays by in-situ electrokinetic remediation," Soil and Sediment Contamination, vol. 6, no. 4, pp. 391–407, 1997.

32. K. R. Reddy, U. S. Parupudi, S. N. Devulapalli, and C. Y. Xu, "Effects of soil composition on the removal of chromium by electrokinetics," Journal of Hazardous Materials, vol. 55, no. 1–3, pp. 135–158, 1997.

33. K. R. Reddy, S. Chinthamreddy, R. E. Saichek, and T. J. Cutright, "Nutrient amendment for the bioremediation of a chromium-contaminated soil by electrokinetics," Energy Sources, vol. 25, no. 9, pp. 931–943, 2003.

34. L. Hopkinson, A. Cundy, D. Faulkner, A. Hansen, and R. Pollock, "Electrokinetic stabilization of chromium (VI)-contaminated soils, electrochemical remediation technologies for polluted soils," inSediments and Groundwater, pp. 179–193, 2009.

35. D. B. Gent, R. M. Bricka, A. N. Alshawabkeh, S. L. Larson, G. Fabian, and S. Granade, "Bench- and field-scale evaluation of chromium and cadmium extraction by electrokinetics," Journal of Hazardous Materials, vol. 110, no. 1–3, pp. 53–62, 2004.

36. P. R. Buchireddy, R. M. Bricka, and D. B. Gent, "Electrokinetic remediation of wood preservative contaminated soil containing copper, chromium, and arsenic," Journal of Hazardous Materials, vol. 162, no. 1, pp. 490–497, 2009.

37. R. H. Myers, D. C. Montgomery, and C. M. Anderson-Cook, Response Surface Methodology: Process and Product

Optimization Using Designed Experiments, John Wiley & Sons, 2009.

38.   M. A. Bezerra, R. E. Santelli, E. P. Oliveira, L. S. Villar, and L. A. Escaleira, "Response surface methodology (RSM) as a tool for optimization in analytical chemistry," Talanta, vol. 76, no. 5, pp. 965–977, 2008.

39.   M. J. Anderson and P. J. Whitcomb, RSM Simplified: Opitimizing Processes Using Response Surface Methods for Design of Experiments, Productivity Press, 2005.

40.   R. N. Yong, M. Nakano, and R. Pusch, Environmental Soil Properties and Behaviour, CRC Press, New York, NY, USA, 2012.

41.   J. Hamed, Y. B. Acar, and R. J. Gale, "Pb(II) removal from kaolinite by electrokinetics," Journal of geotechnical engineering, vol. 117, no. 2, pp. 241–271, 1991.

42.   M. Pazos, A. Plaza, M. Martín, and M. C. Lobo, "The impact of electrokinetic treatment on a loamy-sand soil properties," Chemical Engineering Journal, vol. 183, pp. 231–237, 2012.

43.   S. H. Kim, H. Y. Han, Y. J. Lee, C. W. Kim, and J. W. Yang, "Effect of electrokinetic remediation on indigenous microbial activity and community within diesel contaminated soil," Science of the Total Environment, vol. 408, no. 16, pp. 3162–3168, 2010.

44.   Q.-Y. Wang, D.-M. Zhou, L. Cang, and T.-R. Sun, "Application of bioassays to evaluate a copper contaminated soil before and after a pilot-scale electrokinetic remediation," Environmental Pollution, vol. 157, no. 2, pp. 410–416, 2009.

45.   S. Lukman, M. H. Essa, N. D. Muʾazu, A. Bukhari, and C. Basheer, "Adsorption and desorption of heavy metals onto natural clay material: influence of initial pH," Journal of Environmental Science and Technology, vol. 6, no. 1, pp. 1–15, 2013.

46.   M. H. Essa and M. A. Al-Zahrani, "Date pits as potential raw materials for the production of activated carbons in Saudi Arabia," International Journal of Applied Environmental

Sciences, vol. 4, no. 1, pp. 47–58, 2009.

47. M. H. Essa, M. A. Al-Zahrani, and T. N. Suresh, "Optimization of activated carbon production from date pits," International Journal of Environmental Engineering, vol. 5, pp. 325–338, 2013.

48. EPA, Method 3050B—Acid Digestion of Sediments, Sludges, and Soils, United States Environmental Protection Agency, 1996.

49. USEPA, Method 7000B—Flame Atomic Absorption Spectrophotometry, United States Environmental Protection Agency, 2007.

50. USEPA, Method 7473—Mercury in Solids and Solutions by Thermal Decomposition, Amalgamation, and Atomic Absorption Spectrophotometry, United States Environmental Protection Agency, 2007.

51. J. P. Gustafsso, "Visual MINTEQ ver. 3.0.," 2010,http://www2.lwr.kth.se/English/OurSoftware/vminteq/index.htm.

52. USEPA, Method 3545: Pressurized Fluid Extraction (PFE), United States Environmental Protection Agency, 1996.

53. USEPA, Method 8270D—Semivolatile Organic Compounds By Gas Chromatography/Mass Spectrometry (GC/MS), United States Environmental Protection Agency, 2007.

54. Start-Ease, Design-Expert: Version 8 Software for Window, Start-Ease, Minneapolis, Minn, USA, 2011.

55. D. Derringer and R. Suich, "Simultaneous optimization of several response variables," Journal of Quality Technology, vol. 12, pp. 214–219, 1980.

56. A. Z. Al-Hamdan and K. R. Reddy, "Transient behavior of heavy metals in soils during electrokinetic remediation," Chemosphere, vol. 71, no. 5, pp. 860–871, 2008.

57. A. T. Yeung and Y. Gu, "A review on techniques to enhance electrochemical remediation of contaminated soils," Journal of Hazardous Materials, vol. 195, pp. 11–29, 2011.

58. K. R. Reddy and S. Chinthamreddy, "Enhanced electrokinetic

remediation of heavy metals in glacial till soils using different electrolyte solutions," Journal of Environmental Engineering, vol. 130, no. 4, pp. 442–455, 2004.

59.   A. Alok, R. P. Tiwari, and R. P. Singh, "Effect of pH of anolyte in electrokinetic remediation of cadmium contaminated soil," International Journal of Engineering Research & Technology, vol. 1, pp. 1–11, 2012.

60.   L. M. Vane and G. M. Zang, "Effect of aqueous phase properties on clay particle zeta potential and electro-osmotic permeability: implications for electro-kinetic soil remediation processes," Journal of Hazardous Materials, vol. 55, no. 1–3, pp. 1–22, 1997.

61.   K. Beddiar, T. Fen-Chong, A. Dupas, Y. Berthaud, and P. Dangla, "Role of pH in electro-osmosis: experimental study on NaCl-water saturated kaolinite," Transport in Porous Media, vol. 61, no. 1, pp. 93–107, 2005.

62.   D. L. Sparks, Environmental Soil Chemistry, Academic Press, New York, NY, USA, 2nd edition, 2003.

63.   A. T. Yeung, "Milestone developments, myths, and future directions of electrokinetic remediation,"Separation and Purification Technology, vol. 79, no. 2, pp. 124–132, 2011.

64.   F. E. Asimakopoulou, I. F. Gonos, and I. A. Stathopulos, "Methodologies for determination of soil ionization gradient," Journal of Electrostatics, vol. 70, no. 5, pp. 457–461, 2012.

65.   S. Pamukcu, "Electrochemical transport and transformations," in Electrochemical Remediation Technologies for Polluted Soils, Sediments and Groundwater, pp. 29–65, John Wiley & Sons, New York, NY, USA, 2009.

66.   K. R. Reddy, R. E. Saichek, K. Maturi, and P. Ala, "Effects of soil moisture and heavy metal concentrations on electrokinetic remediation," Indian Geotechnical Journal, vol. 32, pp. 258–288, 2002.

67.   R. J. Hunter and M. James, "Charge reversal of kaolinite by hydrolyzable metal ions: an electroacoustic study," Clays & Clay Minerals, vol. 40, no. 6, pp. 644–649, 1992.

68. K. Maturi and K. R. Reddy, "Cosolvent-enhanced desorption and transport of heavy metals and organic contaminants in soils during electrokinetic remediation," Water, Air, and Soil Pollution, vol. 189, no. 1–4, pp. 199–211, 2008.

69. J. M. Dzenitis, "Steady state and limiting current in electroremediation of soil," Journal of the Electrochemical Society, vol. 144, no. 4, pp. 1317–1322, 1997.

70. K. R. Reddy and M. R. Karri, "Effect of voltage gradient on integrated electrochemical remediation of contaminant mixtures," Land Contamination and Reclamation, vol. 14, no. 3, pp. 685–698, 2006.

71. M. Pourbaix, Atlas of Electrochemical Equilibria in Aqueous Solutions, 1974.

72. K. R. Reddy, S. Danda, and R. E. Saichek, "Complicating factors of using ethylenediamine tetraacetic acid to enhance electrokinetic remediation of multiple heavy metals in clayey soils," Journal of Environmental Engineering, vol. 130, no. 11, pp. 1357–1366, 2004.

73. Y. B. Acar and A. N. Alshawabkeh, "Principles of electrokinetic remediation," Environmental Science & Technology, vol. 27, no. 13, pp. 2638–2647, 1993.

74. J. K. Mitchell, Fundamentals of Soil Behavior, John Wiley & Sons, 1993.

Chapter 4

Advancements In Development of Chemical-Looping Combustion: A Review

He Fang, Li Haibin, and Zhao Zengli

The Renewable Energy and Gas Hydrate Key Laboratory of Chinese Academy of Sciences, Guangzhou Institute of Energy Conversion, Chinese Academy of Sciences, Guangzhou 510640, China

ABSTRACT

Chemical-looping combustion (CLC) is a novel combustion technology with inherent separation of greenhouse CO_2. Extensive research has been performed on CLC in the last decade with respect to oxygen carrier development, reaction kinetics, reactor design, system efficiencies, and prototype testing. Transition metal oxides, such as Ni, Fe, Cu, and Mn oxides, were reported as reactive species in the oxygen carrier particles.Ni-based oxygen carriers exhibited the best reactivity and stability duringmultiredox cycles. The performance of the oxygen carriers can be improved by changing preparation method or by making mixedoxides. The CLC has been demonstrated successfully in continuously operated prototype reactors based on interconnected fluidizedbed system in the size range of 0.3–50 kW. High fuel conversion rates and almost 100% CO_2 capture efficiencies were obtained. The CLC system with two interconnected fluidized-bed reactors was considered the most suitable reactor design. Development of oxygen carriers with excellent reactivity and stability is still one of the challenges for CLC in the near future. Experiences of building and operating the large-scale CLC systems are needed before this technology is used commercially. Chemical-looping reforming (CLR) and chemical-looping hydrogen (CLH) are novel chemical-looping techniques to produce synthesis gas and hydrogen deserving more attention and research.

INTRODUCTION

On February 2, 2007, the United Nations scientific panel studying climate change declared that the evidence of a warming trend is "unequivocal," and that human activity has "very likely" been the driving force in that change over the last 50 years [1]. According to the Intergovernmental Panel on Climate Change (IPCC) of the United Nations, the observed increase in globally averaged temperatures since the mid-twentieth century is very likely to have occurred due to the increase in anthropogenic greenhouse-gas concentrations that leads to the warming of the earth's surface and lower atmosphere. The greenhouse effect is the phenomenon where water vapor, carbon dioxide, methane, and other atmospheric gases absorb outgoing infrared radiation resulting in the raising of the temperature. In its turn, CO_2 is essentially blamed to be the main factor causing the greenhouse effect because it is the most important anthropogenic greenhouse gas [2]. The concentration of

CO_2 in the atmosphere has risen to a value of ~ 370 ppm today, from the preindustrial value of 280 ppm [3]. Combustion of fossil fuels releases a huge amount of carbon as carbon dioxide into the atmosphere. It was reported that fossil fuels fired power production contributes with one third of the total carbon dioxide release from fuels combustion worldwide [4]. It has been well known that increasing concentration of CO_2 to the atmosphere may affect the climate of the earth. It is generally accepted that a reduction in emissions of greenhouse gases is necessary. At present, there are a number of CO_2 capture processes as follows [5]:

- precombustion, in which the hydrocarbon fuels are decarbonized prior to combustion;
- oxyfuel combustion, which uses pure oxygen separated from air;
- postcombustion separation, which separates CO_2 from the flues gases using different methods.

Most of these techniques have large energy penalty and high costs for separation of CO_2 from the rest of the flue gas components, resulting in a significant decrease of the overall combustion efficiency and as a result in a price increase of the energy because of the cost for CO_2 capture. Chemical-looping combustion (CLC) offers a solution for CO_2 separation without energy penalty.

Chemical-Looping Combustion (CLC)

The CLC uses a solid oxygen carrier to transfer the oxygen from the air to the fuel. The advantage with the technique compared to normal combustion is that CO_2 and H_2O are inherently separated from the other components of the flue gas, namely, N_2 and unreacted O_2, and thus no extra energy is needed for CO_2 separation. The CLC system is composed of two reactors, an air and a fuel reactor, as shown in Figure 1.

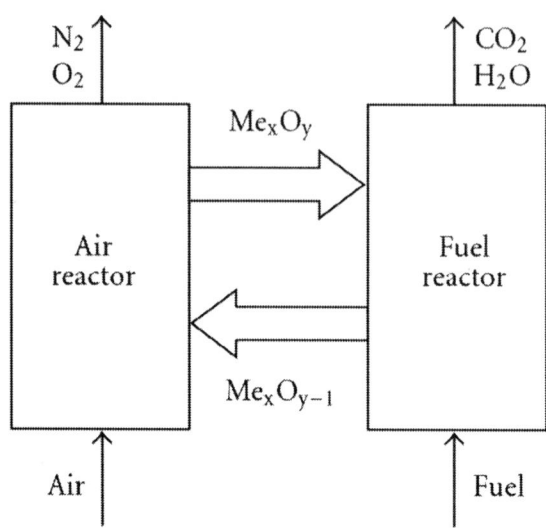

Figure 1: Chemical-looping combustion (CLC).

In CLC, the solid oxygen carrier is circulated between the air and fuel reactors. The fuel is fed into the fuel reactor where it is oxidized by the lattice oxygen of the oxygen carriers according to

$$(2n + m) Me_xO_y + C_nH_{2m}$$

$$\longrightarrow (2n + m) Me_xO_{y-1} + mH_2O + nCO_2, \qquad (1)$$

where M_yO_x is the fully oxidized oxygen carrier and $M_yO_x{-}1$ is the oxygen carrier in the reduced form which could be a metal or a metal oxide with lower oxygen content. The exit stream from the fuel reactor contains only CO_2 and water vapor. The pure CO_2 can be readily recovered by condensing water vapor, eliminating the need of an additional energy for CO_2 separation. The water-free CO_2 can be sequestrated or used for other purpose. Once fuel oxidation completed the reduced metal oxide M_yO_{3-1} is transported to the air reactor where it is reoxidized according the reaction

$$Me_xO_{y-1} + \frac{1}{2}O_2 \longrightarrow Me_xO_y. \qquad (2)$$

The flue gas stream from the air reactor will have a high temperature and contain N_2 and some unreacted O_2. This stream could be expanded through a gas turbine to produce electricity. After energy recovered, these gases can be released to the atmosphere with minimum negative environmental impact. The reaction between the fuel and oxygen carrier in the fuel reactor may be endothermic as well as exothermic depending on the metal oxide used, while the reaction in the air reactor is always exothermic. The CLC does not bring any enthalpy gains, thus, the total heat evolved in these two reactions is the same as that of normal combustion in air. Its main advantage, however, is in the inherent separation of both CO_2 and H_2O from the flue gases. In addition, since air and fuel go through two separated reactors and combustion takes place without a flame, NO_x formation should be avoided [6]. From the point of view of environmental-friendly characterizations, CLC has attracted wide attention and extensive investigation in the past a few years. This work will present an overview of the work which has been conducted about the development and investigation on CLC.

OXYGEN CARRIER DEVELOPMENT

When the CLC was firstly proposed by Richter and Knoche [7], the selection of the oxygen carrier was considered as one of the most important components of the CLC process. The oxygen carrier particles are a cornerstone in the CLC technique. Important properties for oxygen carriers are high reactivity in both reduction by fuel gas and oxidation by oxygen in the air, as well as high resistance to attrition, fragmentation, and agglomeration. Additionally, it is also an advantage if the metal oxide is cheap and environmentally friendly. Briefly, important criteria for a good oxygen carrier are the following:

(i)high reactivity with fuel and air;(ii)low fragmentation and attrition, as well as low tendency for agglomeration;(iii)low production cost and environmentally benign;(iv)be fluidizable and stable under repeated reduction/oxidation cycles at high temperature.

A number of different transition-state metals and their corresponding oxides have been investigated in literature as possible candidates: Cu, Cd, Ni, Mn, Fe, and Co. Figure 2 presents the mass ratio of active oxygen for different systems of metal oxides.

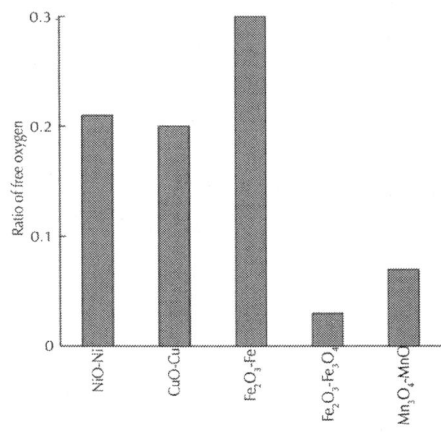

Figure 2: Mass ratio of active oxygen for different metal oxides.

Generally, these metal oxides are combined with an inert which acts as a porous support providing a higher surface area for reaction, as a binder for increasing the mechanical strength and attrition resistance, and additionally, as an ion conductor enhancing the ion permeability in the solid particles [8]. However, Al_2O_3, SiO_2, TiO_2, ZrO_2, $NiAl_2O_4$, and $MgAl_2O_4$ are usually used as the inert binder which was proven to have the ability to increase the reactivity, durability, and fluidizability of the oxygen carrier particles. The inert materials are believed to enhance positive properties among which the most important are to maintain the pore structure inside the particle and inhibit migration of the metals, which could lead to sintering of oxygen carrier particles. In the last decade, a number of researches on oxygen carriers for chemical-looping combustion have been performed. The major affiliations have been Chalmers University of Technology in Göteborg, Sweden, CSIC in Zaragoza,

Spain, Tokyo Institute of Technology in Japan, and Korea Institute of Energy Research.

NiO-Based Oxygen Carriers

Ni-based oxygen carriers can be used at high temperatures of 900–1100°$_C$ in a CLC process with full CH_4 conversion although thermodynamic limitations result in a small amount of CO and H_2 in the outlet gas of the fuel reactor [9]. Pure NiO without doping inert binder used as looping material in CLC has been studied and compared with NiO/YSZ (yttria-stabilized zirconia) by using hydrogen or natural gas as the fuel [6]. As a result, the NiO/YSZ particle permitted extremely fast oxidation compared to the pure NiO sample. The addition of YSZ to NiO gives a high solid diffusivity for the oxide ion especially at high temperature and high porosity turns out to be a crucial feature that YSZ played an important role in increasing the oxidation rate. The performance of NiO/YSZ mixture used as oxygen carrier in CLC was investigated by other researchers [10–14]. It was shown that NiO/YSZ displayed excellent reactivity and regenerability. However, the price of YSZ is higher in comparison to other inerts.

Another inert binder and support material which was used in NiO-based oxygen carriers is Al_2O_3. It was shown that the solid particles of NiO/Al_2O_3 had good reactivity and high mechanical strength [13]. Another advantage of using Al_2O_3 as inert binder is that it is much cheaper than YSZ powder. Therefore, Al_2O_3 has attracted wide attention due to its favorable fluidization properties thermal stability, and low cost [15–18]. It is noted that a part of NiO is converted to form metal aluminum spinel compounds, $NiAl_2O_4$, via solid-state reaction with Al_2O_3 in the course of the sintering process [19]. The $NiAl_2O_4$ is inert or reacts very slowly with fuels or oxygen, therefore, the active metal oxide is added in excess to obtain an NiO/$NiAl_2O_4$ of desired mass ratio to compensate for the loss of nickel as nickel aluminate [20]. $NiAl_2O_4$ is kept even in the cyclic reactions. It plays the role in keeping the mechanical strength of the particle. To inhibit the interaction between NiO and Al_2O_3,

Gayán et al. prepared NiO-based oxygen carriers with modified Al_2O_3 via thermal treatment or chemical deactivation with Mg or Ca oxides [21]. Thermal treatment of y-Al_2O_3 at 1150°C produced the phase transformation to a-Al_2O_3. The chemical pretreatment consisted in precoating the -Al_2O_3 support with MgO or CaO. The precoating was carried out by dry impregnation on y-Al_2O_3 with Mg or Ca nitrate solutions and forming the corresponding aluminates by sintering at high temperature. Ni-based oxygen carriers prepared on -Al_2O_3, $MgAl_2O_4$, or $CaAl_2O_4$ as support showed very high reactivity and high methane combustion selectivity to CO_2 and H_2O because the interaction between the NiO and the support was minimized.

As $NiAl_2O_4$ is formed during the sintering when Al_2O_3 is used as support, a few of researchers have proposed using $NiAl_2O_4$ as a support and binder instead of Al_2O_3 [22–25]. As the results, the $NiO/NiAl_2O_4$ particles displayed good reactivity and selectivity of oxidation of methane to CO_2 and H_2O. The particles showed an increase in mechanical strength, no sign of deactivation, and no change in chemical composition after multicyclic chemical-looping combustion. Linderholm et al. [26] studied the performance of $NiO/NiAl_2O_4$ oxygen carrier in a 10 kW reactor for 160 hours. They found that a conversion of natural gas to CO_2 and H_2O of approximately 99% was accomplished. No decrease in reactivity was seen during the test period. According to the previous investigations, $NiO/NiAl_2O_4$ oxygen carrier was considered as one of the most promising oxygen carriers for CLC due to its good reactivity, cyclic reaction ability, and high bearing temperature of above .900°C.

It was found that the presence of Mg in Ni-Al-O mixed oxides minimizes the sintering of the NiO and stabilizes the Ni^{2+} in the oxide phase [23]. Zafar et al. [27] found that there was no or a limited interaction between the support and the active phase in the course of cycling experiments using a $NiO/MgAl_2O_4$ as oxygen carrier. Therefore, $MgAl_2O_4$ has been considered as a support for NiO to avoid aluminate formation. Johansson et al. [28] tested a chemical-looping combustor of 300 W using an oxygen carrier composed of 60 wt% NiO and 40 wt% $MgAl_2O_4$. It was found that the $NiO/MgAl_2O_4$ particle gave a high conversion of the natural gas.

No methane was detected out in the exit gas out of the fuel reactor, and the fraction of CO varied between 0.5 and 3%. Zafar et al. [29] determined the reactivity of the oxygen carrier composed of 60 wt% NiO with 40 wt % $MgAl_2O_4$. The reactivity was investigated in a TGA at 800-1000°C using 5–20% CH_4 as a fuel gas for reduction and 3–15% O_2 as an oxidizing gas for oxidation. The oxygen carrier showed very high reactivity during reduction and oxidation. The reaction rate was a function of the reacting gas concentration and temperature both in reduction and oxidation reactions. However, it was found that the conversion of particles for the reduction reaction was very low at 800-850°C, which suggested that it may not be feasible to use this oxygen carrier at lower temperature in a CLC system. Johansson et al. found that $NiO/MgAl_2O_4$ demonstrated several advantages at elevated temperatures, that is, higher methane conversion, higher selectivity to reforming, and less tendency for carbon formation compared with $NiO/NiAl_2O_4$ [30]. In addition, several other researchers prepared $NiO/MgAl_2O_4$ oxygen carriers and investigated their performances for CLC [31–33].

A few of researchers have tried to use SiO_2, TiO_2, bentonite, or ZrO_2 as support in the Ni-based oxygen carrier particles for CLC [34, 35]. It was found that the oxygen carriers supported on Al_2O_3 or bentonite produced higher reactivity than those on TiO_2. Moreover, the reactivity of the metal oxide particles increases with increasing temperature and the amount of NiO. Natural titania contained mineral, rutile, was reported to be used as support and binder in nickel-based oxygen carrier [36]. The results revealed that the reactions are fast, as CO_2 is the only compound detected in the outlet gas of the reduction stage. As the reaction proceeds, however, the thermal decomposition of methane appears as a side reaction which competes for methane consumption with the main reaction of the CLC. Zafar et al. [27] observed that the reactivity of NiO/SiO_2 decreased as a function of the cycle number at 950°C but was avoided below 850°C. The same research group reported that NiO/SiO_2 displayed high selectivity toward H_2 during the later stages of reduction in chemical-looping reforming [37].

Generally speaking, the reactivity of the four most studied supported oxygen carriers is in the descending order of NiO > *CuO* > Mn_2O_3 > Fe_2O_3 [38] although this order depends on how the oxide is supported. Therefore, Ni-based oxygen carriers are believed the most promising oxygen carrier candidate for CLC.

CUO-Based Oxygen Carriers

Among the possible metal oxides, CuO has the highest oxygen transport capacity [39]. In the CLC system, the reaction between fuel and metal oxide in the fuel reactor may be endothermic as well as exothermic depending on the oxygen carrier used, while the reaction in the air reactor is always exothermic. When CuO is used as oxygen carrier, however, both the reactions in fuel and air reactors are exothermic. Briefly, Cu-based oxygen carriers have several advantages: (1) CuO has a high oxygen transport capacity, allowing system operating with lower solid flow rates circulating between the fuel reactor and air reactor [40]; (2) the reactions both reduction and oxidation are exothermic avoiding the need of heat supply in the reduction reactor [41]; (3) CuO reduction is favored thermodynamically to reach complete conversion of gaseous hydrocarbon fuels into CO_2 and H_2O [39]; (4) CuO is one of the cheapest materials that can be used for CLC [42]; (5) Cu-based carriers are highly reactive in both reduction and oxidation cycles, which reduces the solids inventory in the system.

de Diego et al. [43] reported that the reaction rate of pure CuO decreased quickly with the increasing number of cycles and after three cycles of reaction the reactivity of the pure CuO was extremely low, reaching conversion of only 10% in more than 20 minutes. Therefore, to obtain better Cu-based oxygen carrier, an inert binder needs to be added into the CuO. They examined the effects of the carrier composition and preparation method on the property of Cu-based oxygen carriers. It was concluded that, to obtain Cu-based oxygen carriers with high reduction and oxidation reaction rates while maintaining their mechanical properties for a high number of successive reduction-oxidation cycles, the only effective preparation method was impregnation on a support. The presence of a binder plays the role as an oxygen-permeable material and as a material to enhance the mechanical strength of the particle for cyclic use and against abrasion. Mattisson et al. [44] prepared CuO/Al_2O_3 oxygen carriers by means of dry impregnation and investigated its reactivity. It was observed that the reduction rate is fast at all temperature in the range 750950∘C for CuO/Al_2O_3 for CuO/Al_2O_3. Minor amounts of CuO were decomposed to Cu_2O during the inert period following the oxidation period at 950∘C. de Diego et al. [45] found that CuO/Al_2O_3 oxygen carriers with a CuO content lower than 10 wt% never agglomerated in the fluidized bed and that the one with a CuO content greater than 20 wt% always agglomerated. In addition, the reactivity of the CuO/Al_2O_3 oxygen carriers, during the reduction and oxidation reactions, was high and not affected by the number of cycles carried out in the fluidized bed. Complete CH_4 conversion to CO_2 and H_2O during most of the reduction period was obtained. Chuang et al. [46] developed Cu-based oxygen carrier and studied the performance for burning solid fuels using CLC. It was found that carriers made by mechanical mixing and wet-impregnation were rejected, because they agglomerated and exhibited low reactivity. Co-precipitated carriers, however, did not agglomerate and showed a high reactivity after 18 cycles of operation. The effect of the operating conditions, such as oxygen carrier-to-fuel ratio, fuel gas velocity, oxygen carrier particle size, and fuel reactor temperature, on fuel conversion was analyzed working with a CuO/Al_2O_3 oxygen carrier prepared by dry impregnation in a 10 KW_{th} pilot fluidized-bed reactor [47]. It was found that the most important parameter affecting the CH_4 conversion was the oxygen carrier-to-fuel ratio. Complete methane conversion, without CO or H_2 emissions, was obtained with this oxygen carrier working at 800°C and oxygen carrier-to-fuel ratios of >1.4.

SiO_2 was studied as well as an oxygen carrier support material for copper-based oxygen carriers. Corbella et al. [48] studied the performance of a copper oxide silica-supported oxygen carrier in a 20-cycle test of chemical-looping of methane in a fixed-bed reactor at 800°C and atmospheric pressure. It was revealed that the reduction reaction rate is fast and highly selective to CO_2 formation, and CO emissions are very low, only yielded at the end of the reduction stage. In the 20-cycle test neither performance decay nor mechanical degradation of the oxygen carrier has been observed. Son et al. [49] found that the oxygen particles supported on SiO_2 exhibit worse reactivity than those on Al_2O_3. In addition, ZrO_2, TiO_2, and bentonite were also studied as the inert support for copper-based oxygen carrier [50–52].

Fe_2O_3-Based Oxygen Carriers

High reactivities in both reduction by fuel and oxidation by oxygen in the air, as well as high resistance to attrition, fragmentation, and agglomeration are the important properties for oxygen carriers. In addition, it is also an advantage if the oxygen carrier is inexpensive and environmentally friendly. Advantages with iron-based oxygen carriers are the environmental compatibility of its oxides magnetite (Fe_3O_4), hematite (Fe_2O_3), and wustite (FeO) and the lower price of iron oxides than other oxides such as nickel oxide (NiO) and copper oxide (CuO) [53]. In view of the availability, low price, as well as environmental safe of iron oxides, has attracted wide attention as oxygen carriers for using in CLC. It was confirmed that the reduction kinetics from hematite to magnetite ($Fe_2O_3 \rightarrow Fe_3O_4$) is the fastest step with the subsequent steps, the steps of magnetite to ferrous oxide ($Fe_3O_4 \rightarrow FeO$) and ferrous oxide to iron (FeO $\rightarrow$ Fe) being much slower. The subsequent steps were considered to be feasible for the chemical-looping reforming (CLR).

Pure Fe_2O_3 was studied as oxygen carrier at 720800°C and demonstrated excellent chemical stability and no loss of activity with cyclic redox. However, the same particles began to agglomerate at 900°C though the agglomeration rate was slow [54]. Previous researchers found that the agglomeration and breakage of the particles could be avoided by adding Al_2O_3 into the particles [55]. Ishida et al. [56] prepared Fe_2O_3/Al_2O_3 composite particles and evaluated their applicability as solid looping materials of the chemical-looping combustor. They found that two solid solutions, hematite (ss) and corundum (ss), were formed in Fe_2O_3/Al_2O_3composite particles at temperatures above 1000°C. The mechanical strength of the Fe_2O_3/Al_2O_3 particles was improved by increasing the content of corundum (ss). Cho et al. [57] compared the iron-, nickel-, copper-, and manganese-based oxygen carriers for chemical-looping combustion. It was observed that oxygen carriers based on nickel, copper, and iron showed high reactivity enough to be feasible for CLC system. However, samples of the Fe_2O_3/Al_2O_3 showed signs of agglomeration. The same research group developed Fe_2O_3-based oxygen carriers together with various inerts such as Al_2O_3, ZrO_2, TiO_2, and $MgAl_2O_4$ finding that Fe_2O_3/Al_2O_3exhibited good reactivity [58]. In our previous work, Fe-based oxygen carrier composed of 80 wt% Fe_2O_3 with 20 wt% Al_2O_3 has been prepared by impregnation methods [59]. The Fe_2O_3/Al_2O_3 oxygen carrier showed good reactivity in 20-cycle redox tests in a TGA reactor. However, 85% of the CH_4 was converted to CO_2 and H_2O during most of the reduction periods with minor formation of CO and H_2. Abad et al. [60] investigated the performance of iron-based oxygen carrier in a continuously operating laboratory CLC unit, consisting of two interconnected fluidized beds using natural gas or syngas as fuel. The combustion of fuel gas was stable during the operation of the reactor. The combustion efficiencies of syngas and natural gas reached 99% and up to 94%, respectively. The reactivity and the crushing strength of the oxygen carrier particles were not affected significantly during operation. Agglomeration and carbon deposition were not observed and no mass loss of the solids inventory was detected. It is clear that attrition and agglomeration of Fe_2O_3/Al_2O_3 oxygen carrier can be controlled at very low level if it is operated under proper conditions. Therefore, Fe_2O_3/Al_2O_3 oxygen carrier is one of the promising candidates for CLC.

Similar to Ni- and Cu-based oxygen carriers, other materials, such as YSZ, $MgAl_2O_4$, TiO_2, and SiO_2, were reported being used as binder and support in the oxygen carrier particles. Johansson et al. [61] suggested that the Fe_2O_3/$MgAl_2O_4$ oxygen carrier showed best reactivity among the iron oxides supported on six different inert materials. As for this kind of oxygen carrier, the one containing 60 wt% Fe_2O_3 and 40 wt% $MgAl_2O_4$ sintered at 1100°C exhibited reasonable crushing strength and resistance toward agglomeration and fragmentation. Leion et al. [62] studied the chemical-looping combustion of petroleum coke using $Fe_2O_3/MgAl_2O_4$ as oxygen carrier. It was revealed that the particles $Fe_2O_3/MgAl_2O_4$ reacted rapidly with intermediate gasification products such as CO and H_2. Therefore, the presence of an oxygen carrier enhanced the gasification of petroleum coke. TiO_2 was also proposed being used as inert and support of the iron-base oxygen carriers. Corbella and Palacios [63] found that the available oxygen of Fe_2O_3/TiO_2 particles for methane combustion in the reduction stage was lower than expected due to the active phase interacts with the support forming $FeTiO_3$ ilmenite. Similar results were observed for Fe_2O_3/SiO_2 oxygen carrier, the formation of Fe_2SiO_4 reduced the reactivity of the Fe_2O_3/TiO_2 particles [37].

Natural iron ores, such as hematite and ilmenite, have also considered as oxygen carriers for CLC, especially for solid fuels CLC. The work in [64] reported the feasibility of using ilmenite as oxygen carrier in CLC. It was found that ilmenite is an attractive and inexpensive oxygen carrier for CLC. The ilmenite particles showed no decrease in reactivity after 37 cycles of redox in a laboratory fluidized-bed reactor system. Ilmenite gave high conversion of CO and moderate conversion of CH_4. Berguerand and Lyngfelt [65] investigated the CLC process of petroleum coke in a 10 kW$_{th}$ chemical-looping combustor. The petroleum conversion reached higher figures of 66 to 78%. Low loss of noncombustible fines from the system indicated very low attrition of the ilmenite particles. The CO_2 capture ranged from 60 to 75%. Natural hematite was also studied being used as oxygen carrier in CLC with CH_4 and air at 950°C [66]. It was found that the majority of CH_4 was converted to CO_2 with some small formation of CO during the reduction period. For the first reduction period the degree of methane conversion was about 62%. Clear breakage on the surface of the reacted particles was observed due to the chemical reactions occurring.

Mn_3O_4-Based Oxygen Carriers

Nickel-, copper-, and iron-oxides have been widely investigated in previous literatures as possible candidates for CLC. Manganese oxides, however, have been reported limitedly being used as oxygen carriers of CLC. Adánez et al. [8] developed manganese oxides on five different inert materials, such as Al_2O_3, Sepiolite, SiO_2, TiO_2, and ZrO_2. In their study, Al_2O_3, Sepiolite, SiO_2, and TiO_2 were found being unsuitable as support for manganese-based oxygen carriers. ZrO_2 was found to be the best inert material for manganese oxides in the view of reactivity and strength. Mattisson et al. [44] prepared Mn_3O_4/Al_2O_3 oxygen carrier and investigated its reactivity. It was found that Mn_3O_4/Al_2O_3 appeared poor reactivity mostly due to the formation of $MnAl_2O_4$ during sintering, which does not react with the fuels and oxygen. Cho et al. [57] also studied the performance of Mn_2O_4 supported on Al_2O_3. Again the reactivity was found to be poor because of the same reason. Johansson et al. [67] synthesized manganese-based oxygen carriers together with pure ZrO_2 and ZrO_2 stabilized with CaO, MgO, and CeO_2. They found that almost all investigated particles exhibited high reactivity and limited physical changes during the cyclic reactions. The effect of the number of cyclic redox reactions on the reactivity of the all four samples was not clear. The one stabilized with MgO showed the highest reactivity among the four kinds of oxygen carriers. In a later study, the same research group investigated the feasibility of the use of an Mn-based oxygen carrier supported on zirconia stabilized with magnesium in a 300 W continuously operating reactor system [68]. As a result, Mn_3O_4/Mg-ZrO_2 particles sintered at 1150°C showed a good reactivity for CLC. Very high efficiencies of >0.999 were obtained for syngas combustion at all temperatures in the range of 800950°C. Agglomeration and attrition rate were not obviously observed. The same authors investigated the reduction and oxidation kinetics of Mn_3O_4/Mg-ZrO_2 oxygen carrier particles for CLC in a subsequent work [69]. The order of reaction was 1 with respect to CH_4 and 0.65 with respect to O_2. The activation energy for the reduction and oxidation reactions were 119 and

19 kJ/mol, respectively. The oxygen carrier particles showed high reactivity during both reduction and oxidation at all investigated temperatures of 800950°C. The oxygen carrier was reduced to MnO in the course of reduction and oxidized to Mn_3O_4 during oxidation at all conditions. It is clear that ZrO_2 stabilized by MgO is a good support and inert binder for Mn-based oxygen carriers.

Other Oxygen Carriers

Beside the aforementioned oxygen carriers, other kinds of materials have been reported being used as oxygen carriers for CLC, such as perovskite, $CaSO_4$, as well as mixed-metal oxides. Actually, perovskites are mixed oxides with the general formula $ABO(3 + \delta)$, where A is usually a ion of a rare earth metal, an alkali metal, or an alkaline earth metal B is a transition metal ion. Both A and B can be partially substituted, leading to a wide variety of compositions of general formula $A_{1-x}A_2B_{1-y}B_yO_{(3+\delta)}$, characterized by structural and electronic defects owing to their nonstoichiometry [70]. The -factor in perovskites can be reduced or increased by changing factors in the conditions such as temperature or O_2 partial pressure. This characterization makes perovskites potentially to be used as oxygen carriers for CLC. Readman et al. [71] studied the feasibility of $La_{0.8}Sr_{0.2}Co_{0.2}Fe_{0.8}$ as a potential oxygen carrier in a chemical-looping reactor. The results suggested that $La_{0.8}Sr_{0.2}Co_{0.2}Fe_{0.8}$ has the redox properties required for chemical looping. Reduction and reoxidation of the perovskite takes place quickly enough for CLC. However, it displayed a low oxygen carrying capacity. Rydén et al. [72] reported the use of $La_xSr_{2-x}Fe_yCo_{1-y}O_{3-\delta}$ as oxygen carriers for CLC or chemical-looping reforming (CLR). They found that $La_{0.5}Sr_{0.5}Fe_{0.5}Co_{0.5}$ has properties that make high conversion of CH_4 into H_2O and CO_2 in the beginning of each reduction period. Substituting La for Sr could increase the oxygen carrying capacity of the perovskite, but reduced the reactivity with CH_4.

The oxygen carriers of individual metal oxides have their own advantages and drawbacks. For example, Ni-based oxygen carriers allow working at high temperature in CLC with high CH_4 conversion, but the drawbacks are the higher carbon formation rate as well as partial oxidation of CH_4 to CO and H_2. On the other hand, Cu-based oxygen carriers give complete fuel combustion to CO_2 and H_2O, but the operating temperature is lower because of the low melting point of the Cu. Mixed-metal oxides may integrate the advantages of each sole oxides as well as partially avoid their disadvantages. Jin et al. in [14] reported that double metal oxides of CoO-NiO/YSZ provided a better performance with good reactivity, complete avoidance of carbon deposition, and significant regenerability for repeated cycles of reduction and oxidation than those of individual metal oxides of NiO/YSZ and CoO/YSZ. Adánez et al. [73] prepared mixed Ni-Cu oxides and examined their performance for CLC of methane. It was revealed that the presence of CuO in the Ni-Cu oxygen carriers gives the full conversion of CH_4 to CO_2 and H_2O with zero CO and H_2 emissions. Additionally, the presence of NiO in the Ni-Cu oxygen carriers allows the particles operating at high temperature (950°C). Hossain and de Lasa [74] developed a bimetallic Co-Ni/Al_2O_3 oxygen carrier for a fluidized bed chemical-looping combustion process. The Co-Ni/Al_2O_3 particles displayed excellent reactivity and stability. They confirmed that the addition of Co in the double metallic Co-Ni/Al_2O_3 particles influences the state of the surface minimizing the formation of nickel aluminate, which was believed to be contributing to the inferior reactivity of Ni-based oxygen carriers. Besides, the addition of Co inhibits metal particle agglomeration during the cyclic redox processes. The activation energy for Co-Ni/Al_2O_3 reduction was found to be less than that of unpromoted Ni/Al_2O_3 samples. This suggested that doping with Co decreases the metal-support interaction and the binding energies between the metals and the fuel molecules [53]. Johansson et al. [75] investigated the synergy effect by using mixed oxides of iron and nickel in combustion of CH_4 in a CLC reactor. It was shown that the mixed-oxide system of 3% nickel oxides in 97% iron oxides gives almost two times as much CO_2 per time unit in comparison to the sum of CO_2 when the oxides were tested separately. They believed that the presence of metallic Ni could firstly catalyze the methane into CO and H_2, which then reacts with iron oxide at a considerably higher rate than methane. Recently, $CaSO_4$ was considered as oxygen carrier in CLC. Song et al. [76] studied the CLC of methane with $CaSO_4$ oxygen carrier in a fixed bed reactor. The results showed that the $CaSO_4$ oxygen carrier has a good reduction reactivity and stability in a long-term redox test at 950°C. The CH_4 conversion rate was found as a function of temperature, gas flow rate, mass of sample loaded, and particle size of oxygen carrier. High temperatures lead to a high CH_4 conversion into CO_2, however, a significant SO_2 formation was observed at higher temperatures. In a later work, the same researchers investigated the cyclic performance of a $CaSO_4$-based oxygen carrier in alternating reducing simulated coal gas and oxidizing conditions [77]. It was found that a high concentration of CO_2 could be obtained in the reduction. The coal gas conversions and CO_2 yield initially increased and finally decreased during the reduction in the multicycle tests. Similar to their previous work, the formation of SO_2 and H_2S during the cyclic tests was found to be responsible for the decay of the reactivity of the $CaSO_4$ oxygen carrier. Therefore, decomposition at high-temperature condition is a major problem for $CaSO_4$ when it is used as oxygen carrier for CLC. Decomposition behaviors of $CaSO_4$ in chemical-looping combustion were studied by Tian et al. [78]. They claimed that the most likely mechanism function in the decomposition of $CaSO_4$ is characterized by the Avrami-Erofeev equation. The decomposition reaction is dominated by the nucleation rate. The work in [76] represented that the release of SO_2 and H_2S can be avoided by optimization of the operating conditions. Addition of a minor amount of fresh limestone into the system of CLC can capture the SO_2, and the products of $CaSO_4$ and CaS can be used as oxygen carriers later [79].

CLC REACTOR DESIGN

Prior to the year 2001, most of the research work on CLC focused on the development of oxygen carriers and on system studies, with limited information about how the CLC reactors could be designed

[80]. In CLC process it is required a good contact between fuels and oxygen carriers as well as a flow of solid material between the two reactors. Although there are several options for CLC system designs, it is likely that interconnected fluidized bed reactors are believed the most suitable reactor design. Lyngfelt et al. [4] proposed a circulating system composed of two interconnected fluidized beds, that is, a high-velocity riser and a low-velocity bubbling fluidized bed, see Figure 3. The bed material (oxygen carrier particles) was circulated between the two fluidized beds. In the air reactor (the riser), oxygen is transferred from the combustion air to the oxygen carrier. In the fuel reactor (the low-velocity fluidized bed), oxygen is transferred from the oxygen carrier to the fuel. The authors presented the critical design parameters, such as the solids inventory, recirculation rate of the oxygen carriers, the dimensions of the reactors, as well as the pressure drops. The same authors constructed a 10 kW chemical looping combustor and operated for 100 hours using NiO-based oxygen carriers and natural gas as fuel in 2005 [81]. However, 99.5% of the added methane was converted into CO_2 and H_2O, with minor of CO, H_2, and unreacted methane in the exit stream. There was no detectable leakage between the two reactors. And CO_2 escape from the system via the air reactor was not found. Therefore, almost 100% of the CO_2 is captured in the process.

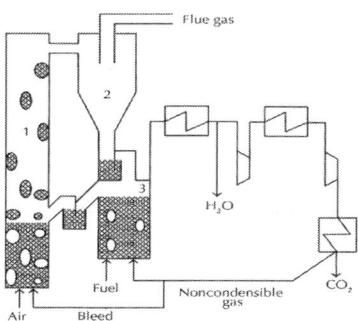

Figure 3: Layout of chemical-looping combustion process, withtwo interconnected fluidized beds. (1) air reactor, (2) cyclone, (3)fuel reactor.

Kronberger et al. [82] designed a 10 kW CLC prototype of a dual-fluidized bed reactor system. Gas velocities and designs were varied, while solids circulation rate and gas leakage between the reactor as well as static pressure balance and residence time distribution of gas and particles were measured. It was observed that the solid circulation rates were sufficient and the gas leakage could be controlled at very low level. In one of their previous work, the same authors developed a 300 W fluidized-bed reactor for CLC [83]. The reactor was tested in various gas velocities and slot design. CLC in the same reactor was continuously operated using Mn-based oxygen carrier [68].

Linderholm et al. [26] examined the CLC process for 160 hours in a 10 kW reactor system with an $NiO/NiAl_2O_4$ oxygen carrier and natural gas as fuel. The prototype consisted of two interconnected fluidized-bed reactors, the fuel and the air reactors, a cyclone to separate solid from gas flow out of the air reactor, and two loop seals, see Figure 4. High fuel conversion to CO_2 and H_2O was achieved. The outlet gas stream out of the fuel reactor mainly contained CO_2 with approximately 0.7% CO, 0.3% CH_4, and 1.3% H_2. The estimated particle life time was 4500 hours. A concentrated stream of CO_2 was obtained when steam was used as fluidization gas in particle locks. Adánez et al. [47] has designed and built a 10 kW pilot plant that is composed of two interconnected bubbling fluidized-bed reactors to demonstrate the CLC technology, see Figure 5. The system was operated for 200 hours, 120 hours of which involved the burning of methane using CuO/Al_2O_3 oxygen carrier. Complete methane conversion was achieved and no deactivation of the oxygen carrier was noticed at 800°C. Similar results were obtained by the same authors [41].

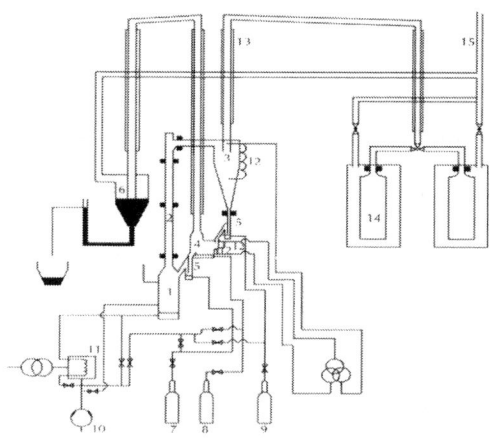

Figure 4: Schematic diagram of the prototype reactor system. (1) air ractor, (2) riser, (3) cyclone, (4) fuel reactor, (5) upper and lower particle locks, (6) water trap, (7) nitrogen (8) natural gas, (9) argon, (10) air, (11) preheater, (12) heating coils (not available for test with nickel-based particles), (13) finned tubes for cooling of gas streams, (14) filters, and (15) connection to chimney.

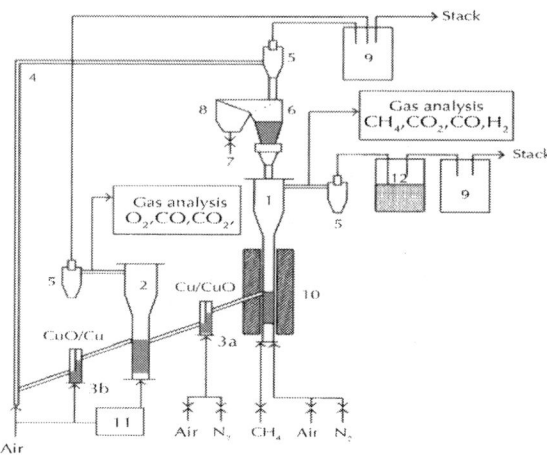

Figure 5: Schematic diagram of the CLC facility of instituto de carboquimica (C.S.I.C.), Spain. (1) fuel reactor, (2) air reactor, (3) loop

seals, (4) riser, (5) cyclone, (6) solid reservoir, (7) solids valve, (8) diverting solid valve, (9) filters, (10) oven, (11) air preheater, and (12) water condenser.

Son and Kim [34] built an annular shape circulating fluidized-bed (CFB) reactor with double loops for investigation into CLC. They tested the CLC of CH_4 using mixed NiO-Fe_2O_3 particles supported on bentonite at 850°C. Full conversion of CH_4 to CO_2 and H_2O was achieved with very small amount of CO and no H_2 emission was detected. Recently, a number of researchers started to investigate the feasibility of CLC for solid fuel. Berguerand and Lyngfelt [84] designed a 10 kW chemical-looping combustor for solid fuels, and tested with South African coal. The CLC system for solid fuels is very similar to the ones for gas fuels, which was reported in literature [22, 26] (see Figure 4). To adapt the solid fuels important modifications in the fuel reactor chamber and the inclusion of an additional solids recirculation loop were made. Additionally, steam was used as the fluidizing agent instead of the fuel gas flow. The chemical-looping combustor has been operated with South African coal for 22 hours, with approximately 12 hours of stable CLC conditions. Ilmenite was proved to be a suitable oxygen carrier for the tested solid fuel with good mechanical properties, low fragmentation/attrition, and good reactivity. The CO_2 capture varied between 82.5% and 96.0% for the coal tests.

Recently, a couple of CLC pilot plants at fuel power of 10–140 kW were built and tested with different fuels. Kolbitsch et al. [85, 86] built a 120 kW CLC pilot unit with dual circulating fluidized bed (DCFB) reactor system, and successfully put it into operation. Two different oxygen carriers, that is, ilmenite and a designed Ni-based particle, were tested in the CLC unit. The experimental results with ilmenite and H_2 rich gases as fuel showed very promising results. High gas conversion was observed. The Ni-based oxygen carrier achieved thermodynamic maximum H_2 and CO conversion as well as very high CH_4 conversion when the solids inventory is sufficiently high. The same research group compared the performance of two Ni-based oxygen carriers, that is, $NiO/NiAl_2O_4$ and $NiO/NiMgAl_2O_4$, for chemical looping combustion of natural gas in the same CLC pilot rig on a scale of 140 kW fuel power [87]. Both oxygen carriers showed high reactivity and no carbon formation was observed under any conditions. Linderholm et al. [88] investigated the reactivity and physical characteristics of Ni-based particles in a 10 kW chemical looping combustor composed of two interconnected fluidized-bed reactors with natural gas as fuel. Long-term of fuel operation (1000 hours) were achieved. The combustion efficiency was around 98% and the methane fraction was typically 0.4–1% in the flue gas of fuel reactor. The particle lifetime was estimated up to 33 000 hours based on calculation from the loss of fines. Shen et al. [89] carried out the experiments for CLC of biomass with iron oxide as an oxygen carrier in a 10 kW continuous reactor of interconnected fluidized beds. The results showed that an increase in the fuel reactor temperature produced a higher increase for the CO production from biomass gasification than for the consumption of CO oxidation to CO_2. The transformation of Fe_2O_3 to Fe_3O_4 is the favored step in the process of iron oxide reduction with biomass syngas. Ryu and Jin [90] proposed a conceptual design a 50 kW thermal chemical-looping combustor composed of two interconnected pressurized circulating fluidized beds. They calculated the important parameters such as bed mass, solid circulation flux, and reactor dimension.

CHEMICAL-LOOPING REFORMING (CLR) AND CHEMICAL-LOOPING HYDROGEN (CLH)

The chemical-looping technique can also be employed for methane reforming and production of hydrogen with inherent CO_2 capture. Chemical-looping reforming (CLR) uses the same basic principles as CLC, with the main difference that the target products in CLR are H_2 and CO instead of heat. The basic principles of chemical-looping reforming are illustrated in Figure 6 [91].

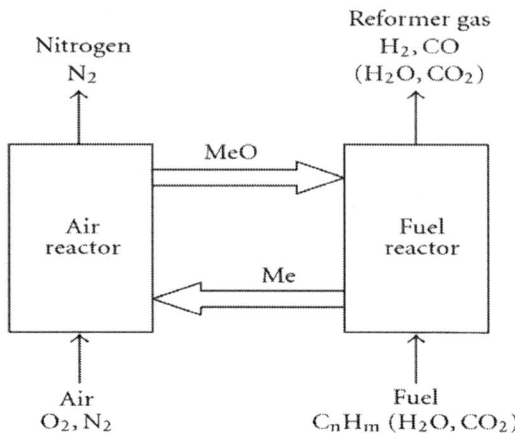

Figure 6: Schematic description of chemical-looping reforming.

In the fuel reactor of CLR process, the fuels are partially oxidized using a solid oxygen carrier to produce synthesis gas, a mix of H_2 and CO, instead of being oxidized into CO_2 and H_2O. Therefore, the ratio of oxygen to fuel is kept low to prevent the fuel from becoming fully oxidized. Pure oxygen production plant which is needed in normal natural gas reforming is avoided. Rydén et al. [91] investigated the CLR of natural gas in a circulating fluidized-bed reactor using Ni-based oxygen carriers. It was found that CLR is a feasible concept for production of synthesis gas and hydrogen. The conversion of natural gas into synthesis gas was 96%–100% depending on oxygen carrier and experimental conditions. The same authors [31] evaluated the feasibility of synthesis gas generation by CLR of natural gas in a continuously operating laboratory reactor using a Ni-based oxygen carrier. As a result, complete conversion of natural gas was achieved and the selectivity towards H_2 and CO was high. Carbon deposition was found for some cases, which was eliminated by adding 25 vol% steam to the natural gas. de Diego et al. [92] studied the synthesis gas generation by chemical-looping reforming of methane using a Ni-based oxygen carrier in a 900 W_{th} CLR pilot plant. The effect of operating conditions, such as fuel reactor temperature, H_2O/CH_4 molar ratio, and solid circulation, on CH_4 conversion and gas product distribution was analyzed. The CH_4

conversion rate reached >98% in all operating conditions. When NiO/CH_4 molar ratio was 1.25, a dry gas product composition of 65 vol% H_2, 25 vol% CO, 9 vol% CO_2, and 1–1.5 vol% CH_4 was obtained in the CLR process. Also, de Diego et al. [93] studied the CLR of CH_4 in a thermogravimetric analyzer (TGA) and in a batch fluidized bed reactor using Ni-based oxygen carriers. The support used to prepare the oxygen carriers exhibited an important effect on the reactivity of the oxygen carriers, on the gas product distribution, and on the carbon deposition. The Ni-based oxygen carrier impregnated on α -Al_2O_3 showed the highest reactivity during the reduction reaction. For all of the tested oxygen carriers, the reduction time without carbon deposition increased with increasing the reduction reaction temperature and the H_2O/CH_4 molar ratio in the feed. The H_2/CO molar ratio in the gas product generated during CH_4 reforming was between 2 and 3 in the reduction period. In a later article, the same researchers presented the experimental results of autothermal CLR of methane in a 900 W_{th} circulating fluidized bed reactor under continuous operation using Ni-based oxygen carriers [94]. It was observed that the CH_4 conversion was very high (>98%) in all operating conditions with both oxygen carriers of 21 wt% NiO supported on γ -Al_2O_3 and 18 wt% NiO on α -Al_2O_3. The oxygen carrier circulation rate, that is, the NiO/CH_4 molar ratio was the most important factor affecting the gas product distribution. Synthesis gas of H_2/CO ratio near 2.6/1 with minor amount of CO_2 and unreacted CH_4 was obtained. Significant changes in the reactivity, surface texture, and solid structure of the oxygen carrier particles were not observed after 50 hours of operation. From these previous literatures, it can be seen clearly that reforming of methane or natural gas by chemical-looping technique is feasible.

Recently, chemical-looping process was proposed being used to produce hydrogen. Chemical-looping hydrogen (CLH) generation, which originates from CLC, actually is a type of water splitting process with a redox of a metal oxide. Similarly to CLC, the CLH system is composed of two reactors: a fuel reactor for burning fuels, and a steam reactor for water decomposition, which is different from the air reactor in the CLC. Fuel is introduced into the fuel reactor to reduce the metal oxide particles, meanwhile the fuel is oxidized into CO_2 and H_2O. The reduced metal oxide is transported to the steam reactor, where it decomposes water to generate H_2. The outlet gas stream from the fuel reactor contains only CO_2 and H_2O at complete conversion of the fuel, while the exit gas from the steam reactor is just H_2 with excess H_2O. Therefore, pure H_2 and CO_2 can be obtained with H_2O condensation without any further separation process [95]. The work in [95] reported that 3.7 L of H_2 was generated through reaction between per kilogram fully reduced copper-based oxide and steam. Go et al. [96] investigated hydrogen production by chemical-looping of methane in a fluidized-bed reactor using iron-based oxygen carrier. It was found that pure hydrogen with free-CO_2 can be obtained from the reaction of $FeO \rightarrow Fe_3O_4$ in steam reactor at 900°C. Additionally, the authors proposed the design basis for the continuous two step steam-methane-reforming (SMR) with double-loop solids circulation system for fuel and steam. Solid fuels, such as coal char, were reported being used as reducing materials in CLR. Yang et al. [97] investigated H_2 production from steam-iron process with reduction of iron oxide by CLC of coal char. Iron oxide was used to oxidize the coal char into CO_2 and H_2O leading to the reduction of Fe_2O_3 to FeO and metallic Fe in the fuel reactor. The reduced iron oxide, FeO and Fe, was oxidized by steam in the steam reactor, while H_2 was generated by water decomposing. Usually, FeO and Fe are oxidized to Fe_3O_4 in the steam reactor. Therefore, an air reactor is needed, where the Fe_3O_4 is oxidized to the original Fe_2O_3 by air. The authors proved that FeO and Fe produced from the Fe_2O_3 reduction of char were feasible for H_2 production through the steam-iron process. The total H_2 produced was 1000 mL of H_2 per gram of tested coal char, where the energy efficiency was 50.2% with respect to the energy ratio of H_2/char.

Similarly, Chiesa et al. [98] proposed a CLH production system with three reactors. The conceptual scheme of the three reactor of CLH is shown in Figure 7. Compared to commercially available technologies, this CLH process shows similar efficiency but much better environmental benign because of the inherent CO_2 separation. They concluded that the CLH system is a promising hydrogen production process deserving substantial research and development activities in the near future. In the steam-iron process to produce hydrogen, deactivation of the iron oxide is one of the problems during the operation. Bleeker et al. [99] considered that the deactivation of the iron oxide is caused by a decrease in the surface area of the oxygen carrier particles. Further, a higher conversion degree of the oxygen carriers in the redox cycles will give a stronger deactivation.

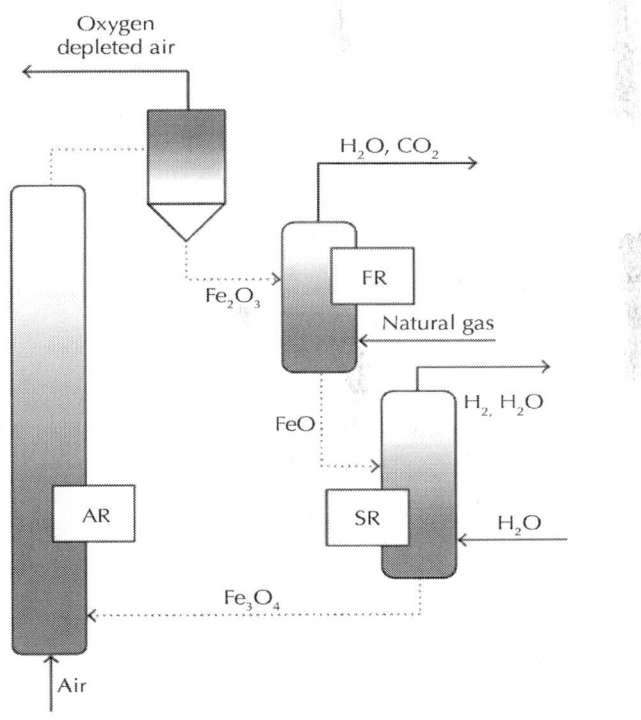

Figure 7: Conceptual scheme of the three-reactor CLH system. AR: air reactor; FR: fuel reactor, SR: steam reactor.

CHEMICAL-LOOPING WITH OXYGEN UNCOUPLING (CLOU)

When CLC process is used to burn solid fuel, there are two approaches operating the combustor to adapt the solid fuels. One option is to introduce the solid fuels directly to the fuel reactor where the gasification of the solid fuels and subsequent reactions with the oxygen carriers will occur simultaneously. Another strategy is to use an oxygen carrier which releases O_2 in the fuel reactor firstly and thereby allowing the fuel to burn with gas-phase oxygen [100]. As for the first method, there is usually a need for an intermediate gasification step of the solid fuels with steam or carbon dioxide to form reactive gaseous compounds which then react with the oxygen carrier particles. Generally, the gasification of solid fuels with H_2O and CO_2 is inherently slow leading to slow overall reaction rates. The second option is referred to as chemical-looping with oxygen uncoupling (CLOU). In the CLOU combustion, the slow gasification of solid fuels with H_2O and CO_2 is avoided, since the fuels reacts directly with gas-phase oxygen. The combustion technique of CLOU involves three steps in two reactors, see Figure 8.

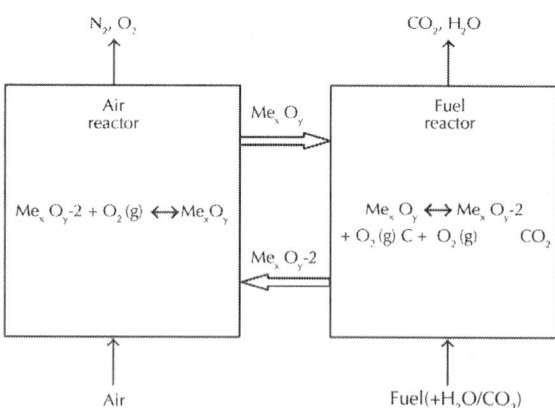

Figure 8: Principal layout of chemical-looping oxygen uncoupling. The oxygen carrier is denoted by Me_xO_y and $MexO_{y-2}$, where Me_xO_y is a met-

al oxide and Me_xO_{y-2} is a metal or metal oxide with lower oxygen content. Here, the fuel is carbon. The fuel reactor could be fluidized using recirculated CO_2 or steam when burning solid fuel [101].

Mattisson et al. [102] presented the chemical-looping with oxygen uncoupling for combustion of petroleum coke. Thermodynamic analysis showed that Mn_2O_3/Mn_3O_4 and CuO/Cu_2O are the most two promising candidates of oxygen carrier for CLOU combustion of solid fuels. The CLOU combustion of petroleum coke was tested in a batch laboratory fluidized-bed reactor and compared to the results of normal CLC. It was found that the reaction rate of petroleum coke was 50 times higher using CLOU than the reaction rate of the same fuel with a Fe_2O_3-based oxygen carrier in normal CLC. In a later work, the same authors [101] studied the CLOU using CuO/ZrO_2 as oxygen carrier and petroleum coke as fuel in a laboratory fluidized bed reactor of quartz. The temperature gave an obvious effect on the reaction rate of petroleum coke. Conversion rates varied between 0.5%/s and 5%/s at a set-point temperature from 895°C to 985°C. These reaction rates are significantly higher than the rates with the same fuel in regular CLC. Therefore, the CLOU is a promising alternative for combustion of solid fuels inherently CO_2 capture.

CARBON FORMATION

Carbon formation may occur during reduction period if carbon-containing fuels are used in CLC. The carbon formation is believed taking place as two mechanisms: pyrolysis of methane and Boudouard reaction (see (3)):

$$CH_4 \longrightarrow C + 2H_2 \quad (pyrolysis),$$

$$2CO \longrightarrow C + CO_2 \quad (Boudouard\ reaction). \tag{3}$$

The pyolysis of methane is thermodynamically favored at high temperature as an endothermic reaction. The Boudouard reaction is exothermic more likely to occur at lower temperatures. Chandel et al. [103] suggested that both of the pyrolysis and Boudouard reactions are slow in CLC process without a catalyst. However, transition metals such as Ni and Fe can act as a catalyst for methane decomposition. Especially, metallic nickel is well known a good catalyst for thermal decomposition of hydrocarbons. This side reaction is undesirable in a CLC process since it increases methane consumption in clear competition with the main reaction in the reduction period. The formed carbon will be burned into CO_2 in oxidation stage resulting in lower efficiency of CO_2 capture of the whole CLC process. Carbon formation using nickel-based oxygen carriers can be decreased by using mixed oxides as oxygen carriers or adding steam into the fuel gas [48]. Other conditions such as oxygen availability of oxygen carriers, fuel conversion, temperature, and pressure could affect the carbon formation. Generally, carbon is prone to be formed at low temperature and small amounts of added oxygen. Here, the oxygen added ratio, c, is defined as the actual amount of O, added with the oxide and/or with steam, over the stoichiometric amount needed for full conversion of the fuel [104]:

$$\varsigma = \frac{n_{O,added}}{n_{O,stoich}}.$$

(4)

Carbon deposition is thermodynamically favorable at a lower oxygen added ratio. Figure 9 shows the carbon formation as a function of temperature and oxygen-added ratio at various pressure independent on the oxygen carrier used when CH_4 is used as fuel in CLC. It is clear that an increased pressure will enhance carbon formation at low temperature, while an increased pressure will reduce carbon formation at high temperature. This is due to the fact that less carbon is formed at higher pressures through methane pyrolysis while more carbon is formed at higher pressures by the Boudouard reaction.

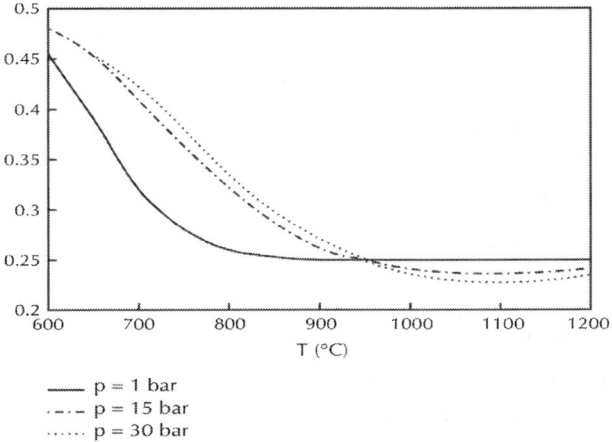

Figure 9: The oxygen added ratio, ç, needed to avoid carbon formation when CH_4 is used as fuel [104].

Addition of water vapor into the fuels before being introduced to the fuel reactor would obviously inhibit the carbon deposition for $NiO/NiAl_2O_4$ particle [13]. The decrease in carbon formation with the addition of steam to methane may be explained with steam reforming and the shift reaction as follows:\

$$CH_4 + H_2O \longrightarrow CO + 3H_2 \quad \text{(Steam reforming)},$$

$$CO + H_2O \longrightarrow CO_2 + H_2 \quad \text{(Shift reaction)}. \tag{5}$$

Apart from above two reactions, carbon formation may be reduced by another reaction as (6) in the presence of steam:

$$C + H_2O \longrightarrow CO + H_2 \quad \text{(Carbon gasification)}. \tag{6}$$

The products H_2 and CO from the aforementioned reactions can be further oxidized by the oxygen carriers to form CO_2 and H_2O. Therefore, addition of steam to the fuels can significantly reduce the carbon formation in CLC. The addition of water vapor resulted in the formation of CO and H_2 as intermediate products. Sometimes, some of the intermediate gases will not be completely oxidized by the oxygen carriers leading to an increase of their concentration in the outlet gas from fuel reactor. Cho et al. [20] investigated the carbon formation conditions on oxygen carriers based on nickel oxide and iron oxide. It was observed that the carbon formation of nickel oxide was considerably affected by the availability of oxygen. Carbon formation was not obvious when sufficient oxygen in the nickel-base oxygen carrier was still available regardless with or without steam addition. For CLR, carbon deposition could be more of a problem since the fuel reactor would be operated under more reducing conditions. For the particles of iron oxide, there was no or very little carbon was formed. Jin et al. [14] reported that carbon deposition of NiO/YSZ oxygen carrier can be completely avoided by doping CoO into the NiO/YSZ. This suggests that carbon formation is also affected by the chemical nature of the oxygen carriers in addition to the operating conditions such as temperature, pressure, gas composition, and particle size.

Basically, carbon formation should not be a problem under the conditions used in a CLC system where a high conversion rate of the fuel is achieved. For a temperature of ,950°C no carbon formation is expected as long as more than one-fourth of the oxygen needed for complete oxidation of CH_4 is supplied [17]. If the CLC process is operated at desired conditions, and a percentage of steam or CO_2 is added into the fuels, carbon formation is considerably inhibited.

CONCLUSIONS

CLC is a novel promising technology for fossil fuels conversion with inherent CO_2 separation. Extensive research has been performed on CLC in the last decade with respect to oxygen carrier development, reaction kinetics, reactor design, system efficiencies, and prototype testing. Ni, Fe, Cu, Mn, and Co oxides are potential candidates for reactive species in the oxygen carrier particles. Ni-based oxygen carriers exhibited the best reactivity and stability during multiredox cycles among these oxides. However, carbon formation is easily generated on Ni-based oxygen carriers when carbon-containing fuels are used in CLC. Doping an impurity such as CoO into the NiO, and adding a proportion of steam into the fuels can considerably inhibit the carbon formation in the reduction period. The performance of other oxygen carriers above mentioned can be improved by changing preparation method or by making mixed oxides.

The CLC process has been demonstrated successfully both in bench scale fixed-bed reactors and in continuously operated prototype reactors based on interconnected fluidized-bed system in the size range of 0.3–140 kW using various types of oxygen carriers and different fuels. In these previous investigations, very high fuel conversion rates up to 100% were achieved in most cases. Therefore, CO_2 capture efficiencies were close to 100%. Although there are several approaches to design the CLC reactor system, it is likely that the interconnected fluidized-bed reactors are considered to be the most suitable reactor design.

In recent years, CLR using the same basic principles as CLC has been proposed to generate synthesis gas by partially oxidizing the natural gas or methane with oxygen carriers in fuel reactor. CLH has been presented to produce pure H_2 with CO_2 capture by means of water splitting process with a redox of an oxygen carrier. Both CLR and CLH have been demonstrated to be feasible in laboratory scale fluidized-bed reactors. In order to adapt to the characterizations of solid fuels, CLOU, which originates from normal CLC, was proposed to burn solid fuels such as coal, petroleum coke, and biomass.

Despite CLC has attracted extensive research in recent years, there are still a number of issues that require further investigation. For example, development of oxygen carriers with excellent reactivity and stability is still one of the challenges for CLC. Construction and operation of large-scale CLC systems is needed before this technology is used commercially. As novel techniques of producing synthesis gas and hydrogen using chemical-looping process, CLR and CLH deserve further research in the near future.

ACKNOWLEDGMENTS

The financial support of National Natural Science Foundation of China (50574046 and 50774038) is gratefully acknowledged. This work was also supported by the Director Foundation of Guangzhou Institute of Energy Conversion (o807z2 and o807rf), Chinese Academy of Sciences.

REFERENCES

1. G. A. Florides and P. Christodoulides, "Global warming and carbon dioxide through sciences,"Environment International, vol. 35, no. 2, pp. 390–401, 2009.

2. P. Cho, T. Mattisson, and A. Lyngfelt, "Defluidization conditions for a fluidized bed of iron oxide-, nickel oxide-, and manganese oxide-containing oxygen carriers for chemical-looping combustion,"Industrial and Engineering Chemistry Research, vol. 45, no. 3, pp. 968–977, 2006.

3. Lyngfelt, B. Leckner, and T. Mattisson, "A fluidized-bed combustion process with inherent CO_2 separation; application of chemical-looping combustion," Chemical Engineering Science, vol. 56, no. 10, pp. 3101–3113, 2001.

4. F. He, H. Wang, and Y.-N. Dai, "Thermodynamic analysis and experimental investigation into nonflame combustion technology (NFCT) with thermal cyclic carrier," Chemical Research in Chinese Universities, vol. 20, no. 5, pp. 612–616, 2004.

5. M. Ishida and H. Jin, "A novel chemical-looping combustor without NO_x formation," Industrial and Engineering Chemistry Research, vol. 35, no. 7, pp. 2469–2472, 1996.

6. H. J. Richter and K. F. Knoche, "Reversibility of combustion processes," in Efficiency and Costing: Second Law Analysis of Processes, vol. 235 of ACS Symposium Series, pp. 71–85, 1983.

7.   J. Adánez, L. F. de Diego, F. García-Labiano, P. Gayán, A. Abad, and J. M. Palacios, "Selection of oxygen carriers for chemical-looping combustion," Energy and Fuels, vol. 18, no. 2, pp. 371–377, 2004.

8.   P. Gayán, L. F. de Diego, F. García-Labiano, J. Adánez, A. Abad, and C. Dueso, "Effect of support on reactivity and selectivity of Ni-based oxygen carriers for chemical-looping combustion," Fuel, vol. 87, no. 12, pp. 2641–2650, 2008.

9.   H. Jin and M. Ishida, "Reactivity study on natural-gas-fueled chemical-looping combustion by a fixed-bed reactor," Industrial and Engineering Chemistry Research, vol. 41, no. 16, pp. 4004–4007, 2002.

10.   M. Ishida, H. Jin, and T. Okamoto, "Kinetic behavior of solid particle in chemical-looping combustion: suppressing carbon deposition in reduction," Energy and Fuels, vol. 12, no. 2, pp. 223–229, 1998.

11.   M. Ishida, H. Jin, and T. Okamoto, "A fundamental study of a new kind of medium material for chemical-looping combustion," Energy and Fuels, vol. 10, no. 4, pp. 958–963, 1996.

12.   H. Jin, T. Okamoto, and M. Ishida, "Development of a novel chemical-looping combustion: synthesis of a solid looping material of NiO/NiAl2O4," Industrial and Engineering Chemistry Research, vol. 38, no. 1, pp. 126–132, 1999.

13.   H. Jin, T. Okamoto, and M. Ishida, "Development of a novel chemical-looping combustion: synthesis of a looping material with a double metal oxide of CoO-NiO," Energy and Fuels, vol. 12, no. 6, pp. 1272–1277, 1998.

14.   M. Ishida, M. Yamamoto, and T. Ohba, "Experimental results of chemical-looping combustion withNiO/NiAl2O4 particle circulation at 1200∘C," Energy Conversion and Management, vol. 43, no. 9–12, pp. 1469–1478, 2002.

15.   H. Zhao, L. Liu, B. Wang, D. Xu, L. Jiang, and C. Zheng, "Sol-gel-derived NiO/NiAl2O4 oxygen carriers for chemical-looping combustion by coal char," Energy and Fuels, vol. 22,

no. 2, pp. 898–905, 2008.

16.   T. Mattisson, M. Johansson, and A. Lyngfelt, "The use of NiO as an oxygen carrier in chemical-looping combustion," Fuel, vol. 85, no. 5-6, pp. 736–747, 2006.

17.   J. E. Readman, A. Olafsen, J. B. Smith, and R. Blom, "Chemical looping combustion usingNiO/NiAl2O4: mechanisms and kinetics of reduction—oxidation (Red-Ox) reactions from in situ powder X-ray diffraction and thermogravimetry expirements," Energy and Fuels, vol. 20, no. 4, pp. 1382–1387, 2006.

18.   P. Cho, T. Mattisson, and A. Lyngfelt, "Defluidization conditions for a fluidized bed of iron oxide-, nickel oxide-, and manganese oxide-containing oxygen carriers for chemical-looping combustion,"Industrial and Engineering Chemistry Research, vol. 45, no. 3, pp. 968–977, 2006.

19.   P. Cho, T. Mattisson, and A. Lyngfelt, "Carbon formation on nickel and iron oxide-containing oxygen carriers for chemical-looping combustion," Industrial and Engineering Chemistry Research, vol. 44, no. 4, pp. 668–676, 2005.

20.   P. Gayán, L. F. de Diego, F. García-Labiano, J. Adánez, A. Abad, and C. Dueso, "Effect of support on reactivity and selectivity of Ni-based oxygen carriers for chemical-looping combustion," Fuel, vol. 87, no. 12, pp. 2641–2650, 2008.

21.   M. Johansson, T. Mattisson, and A. Lyngfelt, "Using of NiO/NiAl2O4 particles in a 10 kW chemical-looping combustor," Industrial & Engineering Chemistry Research, vol. 45, pp. 5911–5919, 2006.

22.   R. Villa, C. Cristiani, G. Groppi, et al., "Ni based mixed oxide materials for CH4 oxidation under redox cycle conditions," Journal of Molecular Catalysis A, vol. 204-205, pp. 637–646, 2003.

23.   J. Wolf, M. Anheden, and J. Yan, "Comparison of nickel- and iron-based oxygen carriers in chemical looping combustion for CO2 capture in power generation," Fuel, vol. 84, no. 7-8, pp. 993–1006, 2005.

24. H.-B. Zhao, L.-M. Liu, D. Xu, C.-G. Zheng, G.-J. Liu, and L.-L. Jiang, "NiO/NiAl2O4 oxygen carriers prepared by sol-gel for chemical-looping combustion fueled by gas," Journal of Fuel Chemistry and Technology, vol. 36, no. 3, pp. 261–266, 2008.

25. C. Linderholm, A. Abad, T. Mattisson, and A. Lyngfelt, "160 h of chemical-looping combustion in a 10 kW reactor system with a NiO-based oxygen carrier," International Journal of Greenhouse Gas Control, vol. 2, no. 4, pp. 520–530, 2008.

26. Q. Zafar, T. Mattisson, and B. Gevert, "Redox investigation of some oxides of transition-state metals Ni, Cu, Fe, and supported on SiO2 and MgAl2O4," Energy and Fuels, vol. 20, no. 1, pp. 34–44, 2006.

27. E. Johansson, T. Mattisson, A. Lyngfelt, and H. Thunman, "A 300 W laboratory reactor system for chemical-looping combustion with particle circulation," Fuel, vol. 85, no. 10-11, pp. 1428–1438, 2006.

28. Q. Zafar, A. Abad, T. Mattisson, and B. Gevert, "Reaction kinetics of freeze-granulated NiO/MgAl2O4oxygen carrier particles for chemical-looping combustion," Energy and Fuels, vol. 21, no. 2, pp. 610–618, 2007.

29. M. Johansson, T. Mattisson, A. Lyngfelt, and A. Abad, "Using continuous and pulse experiments to compare two promising nickel-based oxygen carriers for use in chemical-looping technologies," Fuel, vol. 87, no. 6, pp. 988–1001, 2008.

30. M. Rydén, A. Lyngfelt, and T. Mattisson, "Synthesis gas generation by chemical-looping reforming in a continuously operating laboratory reactor," Fuel, vol. 85, no. 12-13, pp. 1631–1641, 2006.

31. E. Johansson, T. Mattisson, A. Lyngfelt, and H. Thunman, "Combustion of syngas and natural gas in a 300 W chemical-looping combustor," Chemical Engineering Research and Design, vol. 84, no. A9, pp. 819–827, 2006.

32. Lyngfelt, M. Johansson, and T. Mattisson, "Chemical-looping combustion: status of development," inProceedings of the 9th

International Conference on Circulating Fluidized Beds (CFB '08), Hamburg, Germany, May 2008.

33.   S. R. Son and S. D. Kim, "Chemical-looping combustion with NiO and Fe2O3 in a thermobalance and circulating fluidized bed reactor with double loops," Industrial and Engineering Chemistry Research, vol. 45, no. 8, pp. 2689–2696, 2006.

34.   B. M. Corbella, L. F. de Diego, F. García-Labiano, J. Adánez, and J. M. Palacios, "Characterization study and five-cycle tests in a fixed-bed reactor of titania-supported nickel oxide as oxygen carriers for the chemical-looping combustion of methane," Environmental Science and Technology, vol. 39, no. 15, pp. 5796–5803, 2005.

35.   B. M. Corbella, L. F. de Diego, F. García-Labiano, J. Adánez, and J. M. Palacios, "Performance in a fixed-bed reactor of titania-supported nickel oxide as oxygen carriers for the chemical-looping combustion of methane in multicycle tests," Industrial and Engineering Chemistry Research, vol. 45, no. 1, pp. 157–165, 2006.

36.   Q. Zafar, T. Mattisson, and B. Gevert, "Integrated hydrogen and power production with CO2 capture using chemical-looping reforming-redox reactivity of particles of CuO, Mn2O3, NiO, and Fe2O3 usingSiO2 as a support," Industrial and Engineering Chemistry Research, vol. 44, no. 10, pp. 3485–3496, 2005.

37.   M. Johansson, Screening of oxygen-carrier particles based on iron-, manganese-, copper and nickel oxides for use in chemical-looping technologies, Ph.D. thesis, Department of Chemical and Biological Engineering, Environmental Inorganic Chemistry, Chalmers University of Technology, Göteborg, Sweden, 2007.

38.   F. García-Labiano, L. F. de Diego, J. Adánez, A. Abad, and P. Gayán, "Reduction and oxidation kinetics of a copper-based oxygen carrier prepared by impregnation for chemical-looping combustion,"Industrial and Engineering Chemistry Research, vol. 43, no. 26, pp. 8168–8177, 2004.

39. Abad, J. Adánez, F. García-Labiano, L. F. de Diego, P. Gayán, and J. Celaya, "Mapping of the range of operational conditions for Cu-, Fe-, and Ni-based oxygen carriers in chemical-looping combustion,"Chemical Engineering Science, vol. 62, no. 1-2, pp. 533–549, 2007.

40. L. F. de Diego, F. García-Labiano, P. Gayán, J. Celaya, J. M. Palacios, and J. Adánez, "Operation of a 10 kWth chemical-looping combustor during 200 h with a CuO-Al2O3 oxygen carrier," Fuel, vol. 86, no. 7-8, pp. 1036–1045, 2007.

41. M. M. Hossain and H. I. de Lasa, "Chemical-looping combustion (CLC) for inherent CO2 separations-a review," Chemical Engineering Science, vol. 63, no. 18, pp. 4433–4451, 2008.

42. L. F. de Diego, F. García-Labiano, J. Adánez, et al., "Development of Cu-based oxygen carriers for chemical-looping combustion," Fuel, vol. 83, no. 13, pp. 1749–1757, 2004.

43. T. Mattisson, A. Järdnäs, and A. Lyngfelt, "Reactivity of some metal oxides supported on alumina with alternating methane and oxygen—application for chemical-looping combustion," Energy and Fuels, vol. 17, no. 3, pp. 643–651, 2003.

44. L. F. de Diego, P. Gayán, F. García-Labiano, J. Celaya, A. Abad, and J. Adánez, "Impregnated CuO/Al2O3 oxygen carriers for chemical-looping combustion: avoiding fluidized bed agglomeration,"Energy and Fuels, vol. 19, no. 5, pp. 1850–1856, 2005.

45. S. Y. Chuang, J. S. Dennis, A. N. Hayhurst, and S. A. Scott, "Development and performance of Cu-based oxygen carriers for chemical-looping combustion," Combustion and Flame, vol. 154, no. 1-2, pp. 109–121, 2008.

46. J. Adánez, P. Gayán, J. Celaya, L. F. de Diego, F. García-Labiano, and A. Abad, "Chemical looping combustion in a 10 kWth prototype using a CuO/Al2O3 oxygen carrier: effect of operating conditions on methane combustion," Industrial

and Engineering Chemistry Research, vol. 45, no. 17, pp. 6075–6080, 2006.

47.   B. M. Corbella, L. de Diego, F. García-Labiano, J. Adánez, and J. M. Palacios, "Characterization and performance in a multicycle test in a fixed-bed reactor of silica-supported copper oxide as oxygen carrier for chemical-looping combustion of methane," Energy and Fuels, vol. 20, no. 1, pp. 148–154, 2006.

48.   S. R. Son, K. S. Go, and S. D. Kim, "Thermogravimetric analysis of copper oxide for chemical-looping hydrogen generation," Industrial and Engineering Chemistry Research, vol. 48, no. 1, pp. 380–387, 2009.

49.   T. Mattisson, H. Leion, and A. Lyngfelt, "Chemical-looping with oxygen uncoupling using CuO/ZrO2with petroleum coke," Fuel, vol. 88, no. 4, pp. 683–690, 2009.

50.   B. M. Corbella, L. de Diego, F. García, J. Adánez, and J. M. Palacios, "The performance in a fixed bed reactor of copper-based oxides on titania as oxygen carriers for chemicel looping combustion of methane," Energy and Fuels, vol. 19, no. 2, pp. 433–441, 2005.

51.   H. Tian, K. Chaudhari, T. Simonyi, et al., "Chemical-looping combustion of coal-derived synthesis gas over copper oxide oxygen carriers," Energy and Fuels, vol. 22, no. 6, pp. 3744–3755, 2008.

52.   M. M. Hossain, K. E. Sedor, and H. I. de Lasa, "Co-Ni/Al2O3 oxygen carrier for fluidized bed chemical-looping combustion: desorption kinetics and metal-support interaction," Chemical Engineering Science, vol. 62, no. 18–20, pp. 5464–5472, 2007.

53.   R. J. Copeland, G. Alptekin, M. Cesario, and Y. Gershanovich, "Sorbent energy transfer system (SETS) for CO2 separation with high efficiency," in Proceedings of the 27th International Technical Conference on Coal Utilization & Fuel Systems, Clearwater, Fla, USA, 2002.

54.   P. Cho, T. Mattisson, and A. Lyngfelt, "Reactivity of iron oxide

with methane in a laboratory fluidized bed-application of chemical looping combustion," in Proceedings of the 7th International Conference on Fluidized Bed Combustion, p. 599, Niagara Falls, Canada, 2000.

55. M. Ishida, K. Takeshita, K. Suzuki, and T. Ohba, "Application of Fe2O3-Al2O3 composite particles as solid looping material of the chemical-loop combustor," Energy and Fuels, vol. 19, no. 6, pp. 2514–2518, 2005.

56. P. Cho, T. Mattisson, and A. Lyngfelt, "Comparison of iron-, nickel-, copper- and manganese-based oxygen carriers for chemical-looping combustion," Fuel, vol. 83, no. 9, pp. 1215–1225, 2004.

57. T. Mattisson, M. Johansson, and A. Lyngfelt, "Multicycle reduction and oxidation of different types of iron oxide particles-application to chemical-looping combustion," Energy and Fuels, vol. 18, no. 3, pp. 628–637, 2004.

58. F. He, H. Wang, and Y. Dai, "Application of Fe2O3/Al2O3 composite particles as oxygen carrier of chemical looping combustion," Journal of Natural Gas Chemistry, vol. 16, no. 2, pp. 155–161, 2007.

59. Abad, T. Mattisson, A. Lyngfelt, and M. Johansson, "The use of iron oxide as oxygen carrier in a chemical-looping reactor," Fuel, vol. 86, no. 7-8, pp. 1021–1035, 2007.

60. M. Johansson, T. Mattisson, and A. Lyngfelt, "Investigation of Fe2O3 with MgAl2O4 for chemical-looping combustion," Industrial and Engineering Chemistry Research, vol. 43, no. 22, pp. 6978–6987, 2004.

61. H. Leion, T. Mattisson, and A. Lyngfelt, "The use of petroleum coke as fuel in chemical-looping combustion," Fuel, vol. 86, no. 12-13, pp. 1947–1958, 2007.

62. B. M. Corbella and J. M. Palacios, "Titania-supported iron oxide as oxygen carrier for chemical-looping combustion of methane," Fuel, vol. 86, no. 1-2, pp. 113–122, 2007.

63. H. Leion, A. Lyngfelt, M. Johansson, E. Jerndal, and T. Mattisson, "The use of ilmenite as an oxygen carrier in chemical-looping

combustion," Chemical Engineering Research and Design, vol. 86, no. 9, pp. 1017–1026, 2008.

64. N. Berguerand and A. Lyngfelt, "The use of petroleum coke as fuel in a 10 kWth chemical-looping combustor," International Journal of Greenhouse Gas Control, vol. 2, no. 2, pp. 169–179, 2008.

65. T. Mattisson, A. Lyngfelt, and P. Cho, "The use of iron oxide as an oxygen carrier in chemical-looping combustion of methane with inherent separation of CO_2," Fuel, vol. 80, no. 13, pp. 1953–1962, 2001.

66. M. Johansson, T. Mattisson, and A. Lyngfelt, "Investigation of Mn_3O_4 with stabilized ZrO_2 for chemical-looping combustion," Chemical Engineering Research and Design, vol. 84, no. A9, pp. 807–818, 2006.

67. Abad, T. Mattisson, A. Lyngfelt, and M. Rydén, "Chemical-looping combustion in a 300 W continuously operating reactor system using a manganese-based oxygen carrier," Fuel, vol. 85, no. 9, pp. 1174–1185, 2006.

68. Q. Zafar, A. Abad, T. Mattisson, B. Gevert, and M. Strand, "Reduction and oxidation kinetics of Mn_3O_4/Mg-ZrO_2 oxygen carrier particles for chemical-looping combustion," Chemical Engineering Science, vol. 62, no. 23, pp. 6556–6567, 2007.

69. F. He, H. Wang, and Y.-N. Dai, "Preparation and characterization of $La0.8Cu0.2MnO(3\pm\delta)$ perovskite-type catalyst for methane combustion," Transactions of Nonferrous Metals Society of China, vol. 15, no. 3, pp. 691–696, 2005.

70. J. E. Readman, A. Olafsen, Y. Larring, and R. Blom, "$La0.8Sr0.2Co0.2Fe0.8O3-\delta$ as a potential oxygen carrier in a chemical looping type reactor, an in-situ powder X-ray diffraction study," Journal of Materials Chemistry, vol. 15, no. 19, pp. 1931–1937, 2005.

71. M. Rydén, A. Lyngfelt, T. Mattisson, D. Chen, A. Holmen, and E. Bjørgum, "Novel oxygen-carrier materials for chemical-looping combustion and chemical-looping reforming;

LaxSr1−xFeyCo1−yO3−δperovskites and mixed-metal oxides of NiO, Fe2O3 and Mn3O4," International Journal of Greenhouse Gas Control, vol. 2, no. 1, pp. 21–36, 2008.

72. J. Adánez, F. García-Labiano, L. F. de Diego, P. Gayán, J. Celaya, and A. Abad, "Nickel-copper oxygen carriers to reach zero CO and H2 emissions in chemical-looping combustion," Industrial and Engineering Chemistry Research, vol. 45, no. 8, pp. 2617–2625, 2006.

73. M. M. Hossain and H. I. de Lasa, "Reactivity and stability of Co-Ni/Al2O3 oxygen carrier in multicycle CLC," AIChE Journal, vol. 53, no. 7, pp. 1817–1829, 2007.

74. M. Johansson, T. Mattisson, and A. Lyngfelt, "Creating a synergy effect by using mixed oxides of iron- and nickel oxides in the combustion of methane in a chemical-looping combustion reactor," Energy and Fuels, vol. 20, no. 6, pp. 2399–2407, 2006.

75. Q. Song, R. Xiao, Z. Deng, et al., "Chemical-looping combustion of methane with CaSO4 oxygen carrier in a fixed bed reactor," Energy Conversion and Management, vol. 49, no. 11, pp. 3178–3187, 2008.

76. Q. Song, R. Xiao, Z. Deng, W. Zheng, L. Shen, and J. Xiao, "Multicycle study on chemical-looping combustion of simulated coal gas with a CaSO4 oxygen carrier in a fluidized bed reactor," Energy and Fuels, vol. 22, no. 6, pp. 3661–3672, 2008.

77. H. Tian, Q. Guo, and J. Chang, "Investigation into decomposition behavior of CaSO4 in chemical-looping combustion," Energy and Fuels, vol. 22, no. 6, pp. 3915–3921, 2008.

78. L. Shen, M. Zheng, J. Xiao, and R. Xiao, "A mechanistic investigation of a calcium-based oxygen carrier for chemical looping combustion," Combustion and Flame, vol. 154, no. 3, pp. 489–506, 2008.

79. T. Mattisson, Q. Zafar, M. Johansson, and A. Lyngfelt, "Chemical-looping combustion as a new CO2management

technology," in Proceedings of the 1st Reginal Symposium on Carbon Management, Dhahran, Saudi-Arabia, May 2006.

80. Lyngfelt and H. Thunman, "Construction and 100 h of operational experience of a 10-kW chemical-looping combustor," in Carbon Dioxide Capture for Storage in Deep Geologic Formations-Results from the CO_2 Capture Project, vol. 1, pp. 625–645, 2005.

81. Kronberger, A. Lyngfelt, G. Löffler, and H. Hofbauer, "Design and fluid dynamic analysis of a bench-scale combustion system with 2 separation-chemical-looping combustion," Industrial and Engineering Chemistry Research, vol. 44, no. 3, pp. 546–556, 2005.

82. Kronberger, E. Johansson, G. Löffer, T. Mattisson, A. Lyngfelt, and H. Hofbauer, "A two-compartment fluidized bed reactor for 2 capture by chemical-looping combustion," Chemical Engineering and Technology, vol. 27, no. 12, pp. 1318–1326, 2004.

83. N. Berguerand and A. Lyngfelt, "Design and operation of a 10 kWth chemical-looping combustor for solid fuels—testing with South African coal," Fuel, vol. 87, no. 12, pp. 2713–2726, 2008.

84. P. Kolbitsch, T. Pröll, J. Bolhar-Nordenkampf, and H. Hofbauer, "Operating experience with chemical looping combustion in a 120 kW dual circulating fluidized bed (DCFB) unit," Energy Procedia, vol. 1, no. 1, pp. 1465–1472, 2009.

85. J. Bolhàr-Nordenkampf, T. Pröll, P. Kolbitsch, and H. Hofbauer, "Performance of a NiO-based oxygen carrier for chemical looping combustion and reforming in a 120 kW unit," Energy Procedia, vol. 1, no. 1, pp. 19–25, 2009.

86. P. Kolbitsch, J. Bolhàr-Nordenkampf, T. Pröll, and H. Hofbauer, "Comparison of two Ni-based oxygen carriers for chemical looping combustion of natural gas in 140 kW continuous looping operation," Industrial and Engineering Chemistry Research, vol. 48, no. 11, pp. 5542–5547, 2009.

87. C. Linderholm, T. Mattisson, and A. Lyngfelt, "Long-term integrity testing of spray-dried particles in a 10-kW chemical-looping combustor using natural gas as fuel," Fuel, vol. 88, no. 11, pp. 2083–2096, 2009.

88. L. Shen, J. Wu, J. Xiao, Q. Song, and R. Xiao, "Chemical-looping combustion of biomass in a 10 kWthreactor with iron oxide as an oxygen carrier," Energy and Fuels, vol. 23, no. 5, pp. 2498–2505, 2009.

89. H.-J. Ryu and G.-T. Jin, "Conceptual design of 50 kW thermal chemical-looping combustor and analysis of variables," Journal of Energy Engineering, vol. 12, no. 4, pp. 289–301, 2003.

90. M. Rydén, A. Lyngfelt, and T. Mattisson, "Chemical-looping combustion and chemical-looping reforming in a circulating fluidized-bed reactor using Ni-based oxygen carriers," Energy and Fuels, vol. 22, no. 4, pp. 2585–2597, 2008.

91. L. F. de Diego, M. Ortiz, F. García-Labiano, J. Adánez, A. Abad, and P. Gayán, "Synthesis gas generation by chemical-looping reforming using a Ni-based oxygen carrier," Energy Procedia, vol. 1, pp. 3–10, 2009.

92. L. F. de Diego, M. Ortiz, J. Adánez, F. García-Labiano, A. Abad, and P. Gayán, "Synthesis gas generation by chemical-looping reforming in a batch fluidized bed reactor using Ni-based oxygen carriers,"Chemical Engineering Journal, vol. 144, no. 2, pp. 289–298, 2008.

93. L. F. de Diego, M. Ortiz, F. García-Labiano, J. Adánez, A. Abad, and P. Gayán, "Hydrogen production by chemical-looping reforming in a circulating fluidized bed reactor using Ni-based oxygen carriers,"Journal of Power Sources, vol. 192, no. 1, pp. 27–34, 2009.

94. S. R. Son, K. S. Go, and S. D. Kim, "Thermogravimetric analysis of copper oxide for chemical-looping hydrogen generation," Industrial and Engineering Chemistry Research, vol. 48, no. 1, pp. 380–387, 2009.

95. K. S. Go, S. R. Son, S. D. Kim, K. S. Kang, and C. S. Park,

"Hydrogen production from two-step steam methane reforming in a fluidized bed reactor," International Journal of Hydrogen Energy, vol. 34, no. 3, pp. 1301–1309, 2009.

96. J.-B. Yang, N.-S. Cai, and Z.-S. Li, "Hydrogen production from the steam-iron process with direct reduction of iron oxide by chemical looping combustion of coal char," Energy and Fuels, vol. 22, no. 4, pp. 2570–2579, 2008.

97. P. Chiesa, G. Lozza, A. Malandrino, M. Romano, and V. Piccolo, "Three-reactors chemical looping process for hydrogen production," International Journal of Hydrogen Energy, vol. 33, no. 9, pp. 2233–2245, 2008.

98. M. F. Bleeker, H. J. Veringa, and S. R. A. Kersten, "Deactivation of iron oxide used in the steam-iron process to produce hydrogen," Applied Catalysis A, vol. 357, no. 1, pp. 5–17, 2009.

99. H. Leion, Capture of 2 from solid fuels using chemical-looping combustion and chemical-looping with oxygen uncoupling, Ph.D. thesis, Department of Chemical and Biological Engineering, Environmental Inorganic Chemistry, Chalmers University of Technology, Göteborg, Sweden, 2008.

100. T. Mattisson, H. Leion, and A. Lyngfelt, "Chemical-looping with oxygen uncoupling using CuO/ZrO2 with petroleum coke," Fuel, vol. 88, no. 4, pp. 683–690, 2009.

101. T. Mattisson, A. Lyngfelt, and H. Leion, "Chemical-looping with oxygen uncoupling for combustion of solid fuels," International Journal of Greenhouse Gas Control, vol. 3, no. 1, pp. 11–19, 2009.

102. M. K. Chandel, A. Hoteit, and A. Delebarre, "Experimental investigation of some metal oxides for chemical looping combustion in a fluidized bed reactor," Fuel, vol. 88, no. 5, pp. 898–908, 2009.

103. E. Jerndal, T. Mattisson, and A. Lyngfelt, "Thermal analysis of chemical-looping combustion," Chemical Engineering Research and Design, vol. 84, no. 9, pp. 795–806, 2006.

Chapter 5

Gas Migration Along Fault Systems and Through the Vadose Zone in The Latera Caldera (Central Italy): Implications for CO$_2$ Geological Storage

A. Annunziatellis, S.E. Beaubien, S. Bigi,
G. Ciotoli, M. Coltella, and S. Lombardi

Dipartimento di Scienze della Terra, Università di Roma "La Sapienza", Rome, Italy

ABSTRACT

A clear and detailed understanding of gas migration mechanisms from depth to ground surface is fundamental to choose the best locations for CO$_2$ geological storage sites, to engineer them so that they do not leak, and to select the most appropriate monitoring strategy and tools to guarantee public safety. Natural test sites (or "natural analogues") provide the best opportunity to study migration mechanisms, as they incorporate such issues as scale, long-time system evolution, and interacting variables that cannot be adequately addressed with laboratory studies or computer models. To this end the present work examines the migration to surface of deep, naturally produced CO_2 along various buried and exposed faults in the Latera caldera (central Italy) by integrating structural geology and near-surface gas geochemistry surveys. Results show how gas migration is channelled along discrete, high-permeability pathways within the faults, with release typically occurring from spatially restricted gas vents. Size, distribution, and strength of these vents appear to be controlled by the evolution and deformation style of the fault, which is in turn linked to the rheology of the lithological units cut by the fault. As such gas migration can change drastically along strike. Gas migration in the vadose zone around these vents is also discussed, focussing on how the physical–chemical characteristics of various species (CO_2, CH_4, and He) control their spatial distribution and eventual release to the atmosphere.

INTRODUCTION

Research conducted over the last 20 years indicates that the geological storage of anthropogenic CO_2, together with other approaches, is a technically viable option that can help reduce anthropogenic greenhouse gas loading to the atmosphere (IPCC, 2005 and IPCC, 2007). While confidence in CO_2 geological storage is given by the limited number of documented leakages at long-standing sites for CO_2 enhanced oil recovery, acid gas disposal, and natural gas storage (Lewicki et al., 2007a and references therein), and by the positive results obtained from highly monitored CO_2 injection sites like Sleipner (e.g.Chadwick et al., 2006) and Weyburn (Wilson and Monea, 2004), research into specific topics is still needed to ensure the efficiency and safety of this approach.

It is generally agreed that if leakage were to occur from a CO_2 geological storage site the main migration pathways would be along compromised boreholes or gas permeable faults. Whereas the study of industrial sites is needed to better understand the former, natural analogues can be examined to address issues related to the latter. This has been well illustrated by various recent articles (e.g. Baines and Worden, 2004,Beaubien et al., 2005, Holloway et al., 2007, Lewicki et al., 2007a, Pearce, 2004 and Pearce, 2006) that review and compare the results of natural analogues studies as applied to CO_2 geological storage.

Natural analogue sites can be broadly divided into two end-members, those that do not leak and those that do. While well-sealed, natural CO_2 reservoirs can help us to understand the processes that isolate CO_2, and other gases, in the subsurface (e.g. Brennan et al., 2004, Gaus et al., 2004, Le Nindre et al., 2006 and Stevens et al., 2004), it is the leaking CO_2 reservoirs that can be used to study fault-controlled gas migration. These sites, many of which have been leaking CO_2 for hundreds to thousands of years, can be used to address the questions of scale, time, structural and chemical evolution, and natural complexity that cannot be adequately considered in laboratory experiments or computer models.

Large-scale gas leakage in natural systems occurs most commonly in volcanic or geothermal areas, where magma degassing or thermo-metamorphic alteration of carbonates results in the production of large volumes of CO_2. The common occurrence of deep faults associated with these features then provides a permeable pathway along which this CO_2 can migrate towards the atmosphere. A large number of studies have been conducted on these leaking natural analogues to assess the impact of CO_2 on water quality (Wang and Jaffe, 2004), mineral alteration (Stephens and Hering, 2002), and plant and ecosystem health (Pfanz et al., 2007), to quantify the risk to humans associated with the gas released from these structures (Beaubien et al., 2003), and to test innovative monitoring technologies (Anderson and Farrar, 2001). By far most of the papers have been written, however, on the study of the origin, distribution, controlling mechanisms, and mass transfer rates of deep gas to the atmosphere at such well-studied sites as Mammoth Mountain, USA (Gerlach et al., 2001 and Lewicki et al., 2007b), Long Valley, USA (Bergfeld et al., 2006), and the Albani Hills, Italy (Chiodini and Frondini, 2001), to name just a few.

Although many of these papers are put in the context of the structural setting of the area, few are focused directly on the relationship between gas release and the faults and fractures that provide the conduits for upward migration. This approach is more commonly found where soil gas surveys have been specifically conducted across such features (e.g. Azzaro et al., 1998, Baubron et al., 2002, Ciotoli et al., 2007,Fountain and Jacobi, 2000, King et al., 1996, Lewicki and Brantley, 2000 and Lewicki et al., 2003), although the gas migration rates in these studies are typically much lower than that observed above geothermal areas. Studies like these have highlighted the irregular distribution of gas migration along strike, likely due to channelled flow of deep gases along more permeable intervals of the fault and their subsequent release at surface within localised areas aligned along the fault trace.

The issue of fluid movement along faults has also been studied extensively from a structural point of view to better understand the influence of fluids on fault movement and how fault evolution can, in turn, control fluid migration. For example, studies have shown how: (i) fault zones can act as barriers, conduits, or mixed conduit–barrier systems (Antonellini et al., 1994, Caine et al., 1996, Evans et al., 1997 and Sibson, 2000); (ii) structural and hydraulic properties can vary over time and space as a fault evolves (Gibson, 1994,Tellam, 1995 and Wibberley and Shimamoto, 2003); (iii) local- and regional-scale permeability anisotropy can be related to different stages of deformation evolution (e.g. Agosta and Aydin, 2006, Caine and Forster, 1999 and Haynekamp et al., 1999); and (iv) active faults can trigger gas leakage by increasing rock and soil permeability (Uehara and Shimamoto, 2004). Fault permeability models have also been proposed for faults in clastic, crystalline, and, more recently, for carbonate rocks (Agosta and Aydin, 2006, Antonellini et al., 1994, Barton et al., 1995, Bruhn et al., 1994,

Caine and Forster, 1999, Flodin et al., 2005, Rawling et al., 2001 and Sholtz, 1987), although less exists for volcanic terrains (e.g. Gray et al., 2005). Fault permeability has particularly been linked to the lithology of the faulted rocks (Morrow et al., 1984, Zhang and Cox, 2000, Zhang and Tullis, 1998, Zhu and Wong, 1997, Zhu and Wong, 1999 and Zoback and Byerlee, 1975).

It is generally accepted that fault zones consist of two main compartments, a central core surrounded by lateral damage zones (Antonellini et al., 1994, Caine et al., 1996 and Chester et al., 1993). The fault core is the interval in which various mechanical and chemical processes have destroyed the fabric of the host rock (Chester and Logan, 1986 and Sibson, 1977), while the damage zone consists of a wider interval of fractures and smaller faults where the primary fabric is still visible (Cowie and Scholz, 1992). These compartments are then surrounded by weakly deformed host rocks. The thickness of the damage and core zones can vary along a fault plane and, with them, their hydraulic properties (Caine et al., 1996 and Evans et al., 1997). Structural assemblage changes have also been studied as a function of offset variation and distribution along strike (Sholtz, 1987), which is connected to fault growth and the linkage of initially separate fault planes (Morley and Wonganan, 2000, among many others).

The Latera caldera, located in central Italy (Fig. 1 and Fig. 2), is an example of a highly faulted, leaking, natural analogue site. Over the last few years the area above this geothermal field has been studied extensively to better understand the migration, impact, and potential for detection of leaking deep CO_2 (Bateson et al., 2008, Beaubien et al., 2008, Chiodini et al., 2007, Lombardi et al., 2006, Lombardi et al., 2008 and Pettinelli et al., 2008). The present study has used this site to examine how various fault features (such as fault geometry, size, and evolution) influence the migration behaviour of CO_2, as well as the processes in the vadose zone that control the eventual transfer of gas to the atmosphere. Field work involved the close integration of structural geology and near-surface gas geochemistry surveys, as well as examination of historical borehole logs for deep wells (up to 3000 m) drilled in the 1970s for geothermal exploration.

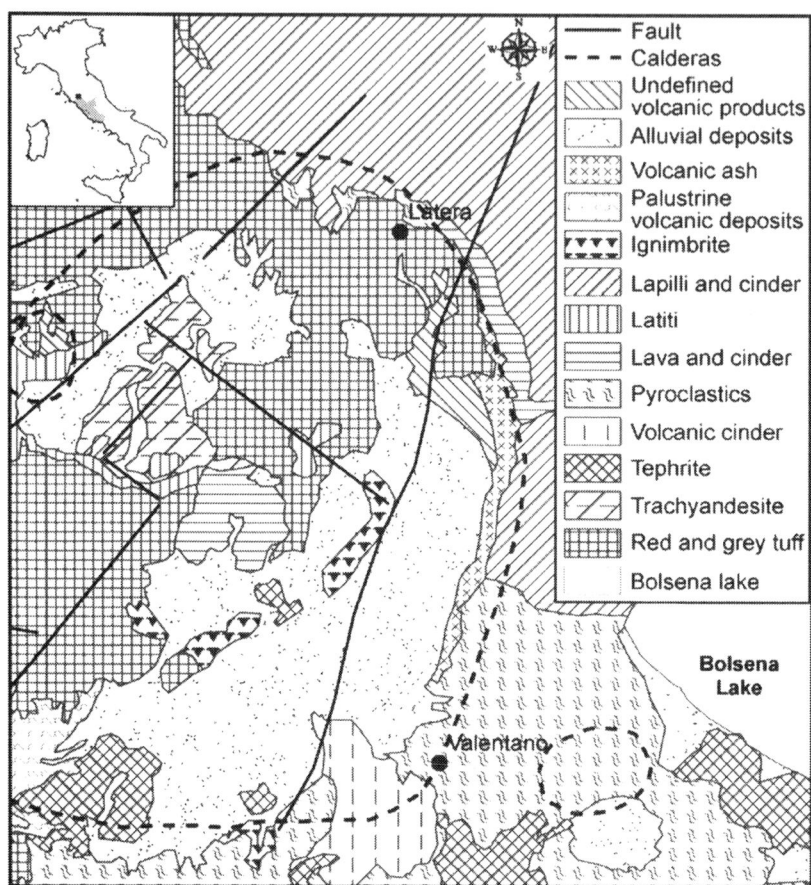

Figure. 1: Geological map of the Latera caldera.

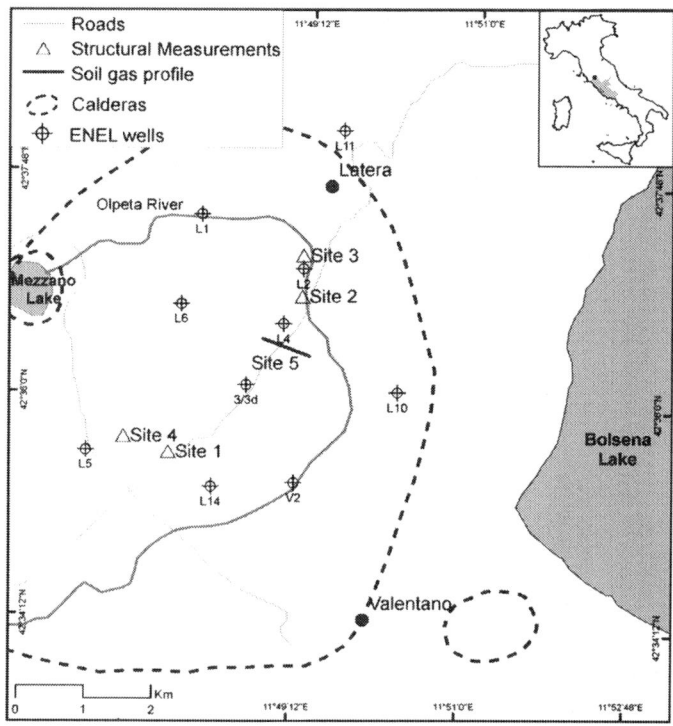

Figure. 2: Map showing the four sites where structural measurements were made (sites 1–4) and the location of a detailed soil gas profile across a number of gas vents (site 5). Three of the sites (1, 3, and 4) are quarries that gave excellent exposure for structural measurements. The location of the deep geothermal exploration wells ("ENEL wells") is given for reference.

GEOLOGICAL AND STRUCTURAL SETTING OF THE LATERA CALDERA

The geology of the west-central Italian peninsula can be thought of as the product of two main tectonic events, a compressive phase during the Eocene to Late Miocene and a subsequent extensional phase from the Late Miocene through the Quaternary (Carmignani and Kligfield, 1990).

The regional substratum of this area consists of metamorphic phyllites and micaschists (see Appendix A for more details), onto which a series of "tectonic units" were placed during the initial compressive period via thrusts that trend primarily N–S. In the area of the Latera caldera, and throughout central Italy, these units are mainly formed by the Mesozoic "Tuscan nappe" and the Cretaceous to Eocene "Ligurian flysch" (Fig. 3). The Tuscan nappe consists primarily of carbonates and some siliciclastic successions; because of high secondary permeability in the carbonates, this unit forms a regional aquifer and locally hosts the geothermal reservoir beneath the Latera caldera. The overlying Ligurian flysch is made up of various units, including impermeable shales and siltstones which act as a regional aquitard and which locally form the caprock for the geothermal reservoir and associated CO_2.

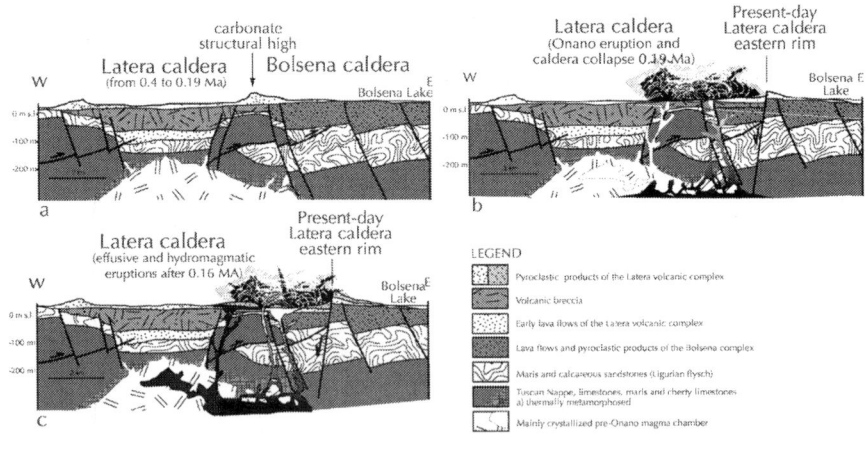

Figure. 3: W–E schematic geological sections illustrating the main steps of the evolution of the Latera caldera; subsurface data after Barberi et al. (1984) and Palladino and Simei (2005). The individual figures refer to the main period of explosive activity from 0.4 to 0.2 Ma (a), the Onano eruption around 0.17 Ma (b), and hydromagmatic eruptions during the later stages of activity (c). See the text and Appendix A for details.

The subsequent extensional regime, which was related to the opening of the Tyrrhenian Sea, resulted in the formation of a series of NW-SE to N–S trending grabens and normal fault systems which dissected the previous compressional fold and thrust belt (Di Filippo et al., 1999 and Funiciello and Parotto, 1978). The extensional phase also caused crustal thinning, resulting in significant volcanic activity during the Quaternary. The Vulsini volcanic district was formed during this period, consisting of the three main eruptive centres of Latera (Fig. 1), Bolsena, and Montefiascone. Volcanism in this area began about 0.8 Ma ago (Evemden and Curtis, 1965 and Nicoletti et al., 1979), with most of the Latera products being dated from 232 to 156 ka (Turbeville, 1992). More details regarding the local geology are given in Appendix A.

At the cessation of Latera volcanic activity, edifice collapse created the caldera via a series of associated normal faults which, combined with thrusts, regional normal faults, and extensive fracturing associated with the volcanic eruptions themselves, became conduits for upwardly migrating hydrothermal fluids and CO_2. Many of these conduits became sealed via secondary mineral precipitation (Cavarretta et al., 1985). This process has made much of the overlying flysch and volcanic units impervious, except where gas is locally escaping, and helped to partially isolate the underlying geothermal reservoir.

The reservoir itself is located within a structural high of Tuscan limestones (Barberi et al., 1984 and Palladino and Simei, 2005) (Fig. 3), a feature which may represent a recumbent fold (Bertrami et al., 1984), an ancient caldera rim (Barberi et al., 1984), or a series of superimposed thrust sheets (this work). This structural high is bounded laterally by sealed faults and thermo-metamorphosed carbonates, below by the metamorphic basement, and above by low permeability Ligurian flysch and volcanic pyroclastics. The reservoir limestones are heavily fractured and folded, then locally re-sealed with a composite hydrothermal assemblage (Cavarretta et al., 1985). Maximum permeability within the reservoir occurs along the major, NNE–SSW-trending axis of the structure (Sabatelli and Mannari, 1995).

The CO_2 that leaks at the surface is likely the product of metamorphic alteration of the reservoir limestones, driven by the water-dominated system and the high heat flow (presently ranging from 170 to 230 °C) linked to the original magma intrusion (Chiodini et al., 1995, Duchi et al., 1992 and Minissale, 2004), although a portion may also come from a deeper mantle component (Chiodini et al., 2007). Because the geothermal system is water-dominated the majority of gas samples collected from the deep wells are dissolved gases that exsolve at the lower pressures at surface. In general, most reservoir waters contain 3–6% "Non Condensable Gases" (NCG) (Sabatelli and Mannari, 1995), with well L2 (Fig. 2) having a representative composition of 98.35% CO_2, 0.05% CH_4, 1.22% H_2S, 0.4% N_2 and trace levels of H_2 (Bertrami et al., 1984). One well (L11; Fig. 2) did intersect a gas-only reservoir having 98% CO_2 (Lombardi et al., 1993). Based on calculations using total surface heat flux, enthalpy of liquid water at reservoir temperatures, and the CO_2 molality in the reservoir water, Gambardella et al. (2004) estimated a total CO_2 production rate of 2.58×10^8 mol CO_2 year^{-1} for the approximately 20 km² surface area of the reservoir.

At the surface the central part of the Latera caldera is characterised by a NE–SW to N–S trending hydrothermally altered area, with silica, alunite, and kaolinite formed in the vicinity of several gas vents and thermal/cold springs (Gianelli and Scandiffio, 1989, Lombardi and Mattias, 1987 and Lombardi et al., 1993). These two directions have also been observed by Buonasorte et al. (1991) in a geomorphological study of the dominant morphological elements located throughout the Vulsini mountains.

STUDY SITE DESCRIPTION

The Latera caldera, located about 110 km to the NW of Rome, Italy (Fig. 2), is a NE–SW trending elliptical depression consisting of a relatively flat central agricultural plain surrounded by tree-covered, 100 m high hills. Except for some local hills formed by resistive volcanic units, the caldera plain is covered by tens of metres of alluvium.

In addition to a regional examination of the Latera caldera, five small sites were chosen for detailed work (Fig. 2); structural measurements were made at sites 1–4, CO_2 flux measurements were performed at site 3, and a horizontal profile of soil gas and CO_2 flux measurements was conducted at site 5. Site 1 is a large, presently active, open-pit kaolinite quarry which has 5–10 m high exposures on its south-western and south-eastern walls; the other two sides of the quarry are lower with poor exposure and extensive vegetation cover. Site 2 is a linear rock escarpment into which a series of caves have been dug; these caves have been blocked off by local authorities due to the accumulation of CO_2 and H_2S coming from fractures in the surrounding rock. Site 3 is an abandoned kaolinite and sulphur mine cut into a hillside; because of the extensive work conducted at this site it is described in more detail below. Site 4 is a very small exposure in thick forest that at one time was likely excavated for potential kaolinite extraction but never developed; the 3 m high exposure is cut into a large hill on the NE side. Finally, site 5 is located on the caldera plain. Here there is no exposure, but instead relatively flat agricultural fields are located on 10–50 m thick alluvial sediments that host an unconfined aquifer at about 4 m depth.

Site 3 was chosen for extensive structural and geochemical surveys due to the excellent exposure of a large fault structure. Quarrying into the hillside at this site created a "U"-shaped feature having 2–10 m high quarry walls on three sides that surround a flat quarry floor. Three mine adits were originally dug into the south-western face of the quarry, first for sulphur until 1950 and then for kaolinite until its closure in the early 1970s. For safety reasons, the tunnel entrances were closed at this time with the same wall material. The fault, which is exposed on this south-western face, strikes across the flat quarry floor where a detailed gas flux survey was conducted. Interviews with ex-employees of the mine indicate that the tunnels went straight into the hillside (i.e. towards the southwest), and thus there should not be any man-made feature beneath the quarry floor (located to the north-east). The quarry floor is devoid of vegetation and covered with a thin (5–30 cm) layer of fine sediments, although the underlying fractured and altered volcanic rocks can be seen in some locations.

METHODOLOGY

Analysis of Brittle Deformation

A preliminary, regional analysis of faults and fractures was first carried out in the Latera caldera. Due to limited outcrop exposure, about 70 fracture and small fault measurements were collected in the four kaolinite and sulphur quarries described above (Fig. 2). All data have been projected on equi-areal Schmidt diagrams (lower hemisphere) and plotted on merged pole-to-plane and density diagrams.

Soil Gas Geochemistry

Soil gas samples were collected using a 6.4 mm diameter, thick-walled, stainless-steel probe pounded in the ground with a co-axial hammer to a depth of about 60–80 cm to avoid the influence of atmospheric air (Hinkle, 1994). All soil gas surveys were conducted during low-precipitation seasons (typically summer to early fall) to minimise any variations induced by different sampling periods. As such it is believed that all the surveys represent the same populations and that they can be combined for statistical and geo-spatial analysis.

Field analyses were conducted for carbon dioxide (CO_2), hydrogen sulphide (H_2S), and hydrogen (H_2) using a gas analyser (Draeger Multiwarn) connected directly to the probe. Samples were also collected from the probe for laboratory analysis by injecting 60 ml of soil gas into a previously evacuated, 25 ml stainless-steel container having a screw top port with a compressible septum. These containers were analysed in the laboratory first for helium (He) on a mass spectrometer (Varian Leak Detector) and then for other gases on two Fisons 8000-series bench gas chromatographs. These gases included CO_2, oxygen + argon (O_2 + Ar), nitrogen (N_2), methane (CH_4), and ethane (C_2H_6).

Two soil gas datasets are presented here. The first consists of a combined dataset formed by various regional (2 samples km^{-2}) and detailed (40–50 samples km^{-2}) soil gas surveys performed throughout the Latera caldera over the last number of years. The total number of sample points in this dataset depends on the gas type (He, 2467; CO_2, 1735; CH_4, 938). The second dataset is formed by 202 points sampled along a detailed, 550 m long horizontal profile crossing a number of gas vents (site 5, Fig. 2). Not all points were sampled for all parameters. Instead, sample spacing along the profile for the field analysis of soil gas CO_2 and CO_2 flux was 2–4 m, while it was 8–12 m for laboratory analysis of soil gas He and CH_4 in low CO_2 areas and every 4 m in anomalous CO_2 areas.

CO_2 Flux Geochemistry

Carbon dioxide flux measurements were made using a closed-circuit accumulation system (e.g.Hutchinson and Livingston, 1993) that was constructed "in-house". The control unit is equipped with a 0–3000 ppm infrared CO_2 detector while the accumulation chamber itself is a small Plexiglas box with an inlet and outlet tube for the IR detector and a 30 cm long, 5 mm diameter pressure equilibration tube to prevent artefacts above high flux points and on windy days (Hutchinson and Mosier, 1981). The accumulation chamber was pressed firmly to the ground to insure a proper seal after the removal of any surface vegetation, and CO_2 concentrations were measured and stored every second for about 100 s. Flux values were calculated using:

$$\phi CO_2 = \left(\frac{d[CO_2]}{dt}\right)\left(\frac{V}{A}\right)\left(\frac{T_0}{T}\right)\left(\frac{P}{P_0}\right)k$$

$$(1)$$

where ($d[CO_2]/dt$) is the first linear slope of the CO_2 concentration increase (ppm/s), V is the chamber volume (m^3), A is the chamber surface area (m^2), T and T_0 are measured and standard temperature

(°K), P and P_0 are measured and standard pressure (kPa), and k is a unit conversion factor (169.71).

A highly detailed grid of CO_2 flux measurements was performed on the floor of the site 3 quarry (Fig. 2) to understand the distribution of gas migration relative to the local structural features. Flux measurements were conducted on a regular grid covering a total area of 50 m × 70 m and having a sample spacing of 3 m × 5 m; samples were collected every 3 m along a series of parallel lines trending N350°E (i.e. perpendicular to the main fault direction) and located 5 m apart. Soil gas samples were not collected here due to the very shallow occurrence of solid rock. CO_2 flux measurements were also conducted at site 5 along the detailed horizontal profile described above.

Gas Geochemistry Data Analysis

Preliminary data analysis involved the calculation of standard statistical parameters to evaluate the basic characteristics of the data, the creation of normal probability plots (NPP) to define statistical populations for each parameter and to choose contour thresholds, and the creation of post and classed post maps to analyse sample and value distribution.

A subsequent geostatistical study consisted of experimental variogram calculations to check spatial continuity of the data values and the presence of anisotropies. As geostatistical analyses provide poor results with non-normal sample distributions, the CO_2 flux and soil gas concentration data were log-transformed to obtain a normal distribution and to reduce the influence of outliers. Once the data spatial continuity was characterised it was modelled using variogram functions that formed the basis for contouring with the kriging method. Finally, cross-validation was used to validate the model parameters used for the estimation. Collected data were processed using the following software: Grapher 6.0 and Surfer 8.0 (Golden Software, Inc.), Statistica 6.0 (StatSoft, Inc.), GS + (Gamma Design Software, LLC), and ArcGIS (ESRI, Inc.).

RESULTS

Regional Study

Surveys were conducted at the caldera scale to better understand the overall structural architecture of the area and how this controls the migration of gas to the surface.

Structural Geology

Extensive structural measurements were collected at four sites located throughout the valley floor (Fig. 2). In general, fractures and fault systems observed at these locations vary locally from a very low density (about one fracture every 150 m) to higher values (about one fracture every 50 cm). Combined pole-to-plane and density plots (Fig. 4) of this data show clusters around N50°E and N10°E, which is in agreement with other measurements made by the authors on the caldera edge and rim.

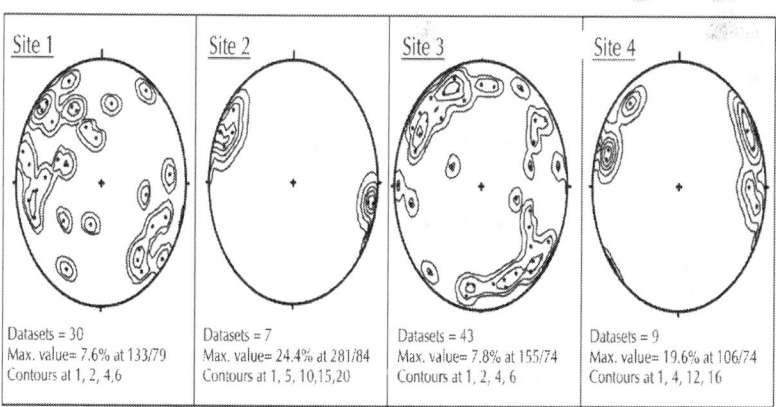

Figure. 4: Fault pole and pole density plots of the faults and fractures measured at each site.

Fractures belonging to the N50°E system have well-defined surfaces that are generally open and filled by shale. The opening is typically mode I (i.e. opening normal to the fracture walls), but several of these fractures belong to mode II, as shown by small pull-apart features along the planes that indicate a normal sense of shear. The N10°E fractures are mode I and are generally associated with fine breccias of a few millimetres. Most of these fractures reach the field surface, with evidence of shallow fluid circulation (black to brown alteration and encrustations, irregular planes) (Fig. 5a).

Figure. 5: Examples of faults and fractures from the Latera caldera area. (a) Fracture with an irregular plane in pyroclastic deposits at the site 1

quarry (Fig. 2); (b) high angle normal fault in lavas altered to kaolinite at the site 3 quarry; (c) small normal fault in pyroclastic deposits at the site 1 quarry.

Small faults have been measured in all the quarries. They have an offset of about 20–40 cm, strike from N20°E to N50°E (Fig. 5b and c), and most show evidence of active fluid circulation in the form of secondary sulphate minerals.

Soil Gas Geochemistry

As mentioned, a number of regional and detailed soil gas surveys throughout the Latera caldera were combined to compare soil gas trends with the results of the structural geology surveys described above.Table 1 summarises various statistical parameters for three gases (CO_2, CH_4, He) of this combined dataset and contrasts them with an Italian-wide dataset, created by the authors over the last 30 years of soil gas sampling, and a sub-set of just-volcanic areas. The combined Latera dataset shows a skewed distribution for CO_2 and CH_4, as indicated by the large difference between mean and median values and by the high standard deviation values (Table 1). In contrast the He distribution is primarily Gaussian, except for a few outliers. When compared with other volcanic areas, and the database as a whole, the Latera dataset has the highest mean and median CO_2 and CH_4 values, while He values are slightly lower.

Table 1: Soil gas statistics for the regional and detailed sites at Latera, compared to the statistics for a pan-Italian database produced by the authors over the last 30 years

	N	Mean	Median	Min	Max	LQ	UQ	0.10%	0.90%	IQR	S.D.
Latera regional study											
He (ppm)	2,467	5.35	5.29	4.24	9.86	5.17	5.42	5.01	5.62	0.25	0.45
CO_2 (%)	1,735	6.74	1.30	0.03	100.00	0.64	4.22	0.34	19.60	3.58	15.09
CH_4 (ppm)	938	26.40	1.09	0.11	1345.34	0.56	2.03	0.33	18.50	1.47	126.92
Sulphur quarry (site 3)											
CO_2 flux (g m^{-2} d^{-1})	267	1700.28	331.52	7.02	49563.63	138.22	1050.27	47.11	3828.30	912.05	4992.67
Horizontal profile (site 5)											
He (ppm)	75	5.38	5.33	4.63	8.13	4.90	5.47	4.80	6.93	0.56	0.82
CO_2 (%)	82	17.68	8.70	0.42	85.92	3.20	19.51	1.47	52.35	16.30	2.38
CH_4 (ppm)	82	57.61	0.32	0.00	1019.04	0.16	0.65	0.16	4.41	0.49	218.37
CO_2 flux (g m^{-2} d^{-1})	147	131.10	5.06	3.25	3569.73	9.05	55.78	5.70	210.26	46.72	426.43
Other Italian volcanic areas											
He (ppm)	15,080	5.45	5.33	1.20	172.96	5.22	5.48	5.04	5.70	0.26	2.29
CO_2 (%)	4,860	3.33	0.88	0.03	100.00	0.41	1.92	0.18	4.81	1.51	10.32
CH_4 (ppm)	3,395	16.61	1.69	0.01	7104.26	1.04	2.39	0.63	3.83	1.35	167.08
Complete Italian database											
He (ppm)	38,060	5.48	5.31	1.20	315.22	5.20	5.48	5.02	5.76	0.28	2.95
CO_2 (%)	16,301	1.93	0.83	0.03	100.00	0.38	1.69	0.16	3.22	1.31	6.09
CH_4 (ppm)	11,945	14.65	1.83	0.01	19396.14	1.16	2.54	0.72	3.41	1.38	263.10

Experimental variograms were calculated for the log-transformed Latera CO_2 dataset using the average distance between samples (about 200 m) as the starting lag. The resultant variogram surface maps (not shown) highlight the presence of two main anisotropy axes acting at different spatial scales and different directions: (i) a small-scale behaviour along a direction of N45°E, with a spatial range of 350 m along the major axis of anisotropy and 150 m along the minor axes of anisotropy; and (ii) a large-scale behaviour along a direction of N20°E, with a spatial range of 800 m along the major axis of anisotropy and of 300 m along the minor axes of anisotropy. These two directions are in good agreement with that observed in the structural results presented above. The resultant equation of the variogam model is = 1.4 Nugget + 1 Sph (350, 150, 45) + 0.6 Sph (800, 300, 20), which was used in the kriging calculations to interpolate/extrapolate values at un-sampled locations. Furthermore, the calculation of the experimental variogram perpendicular to the large-scale anisotropy axis shows repetitive high and low spatial variance, indicating the presence of a phenomenon that repeats itself in space with a range of about 300 m.

The resultant contour map (Fig. 6) shows a number of elongated soil gas CO_2 anomalies that trend from N10°E to N50°E. Values above 2.5% (outlined with a black line). Values above 2.5% were considered anomalous based on the normal probability plot distribution of this dataset.

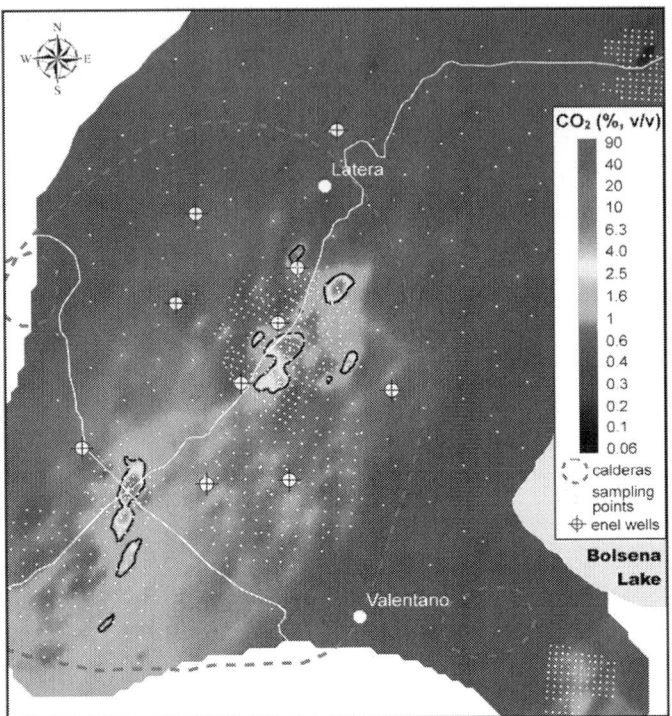

Figure. 6: The contoured distribution of CO_2 soil gas concentrations throughout the Latera caldera area, with anomalous values above 2.5% CO_2 outlined in black. White dots show the sampling locations while the deep geothermal exploration wells are the same as given in Fig. 2.

The results for the detailed profile conducted at site 5 across a number of gas vents in the centre of the Latera caldera are shown in Fig. 7, where soil gas CO_2, CH_4, and He are plotted together with CO_2 flux values. The trend of all four parameters is generally similar, with elevated values observed in the intervals from 0 to 50 m, 200 to 225 m, 300 to 360 m, and 400 to 500 m. These four venting features are well resolved with this detailed profile, compared to the large, "spatially averaged" anomalies that are observed in the regional map of Fig. 6. The sections between these vents clearly show background values that are not influenced by deep leaking CO_2; background values for the profile dataset, defined using normal probability plots, are: CO_2 flux < 22 g m^{-2} d^{-1};

$[CO_2] < 2.5\%$; $[CH_4] < 1.5$ ppm; and $[He] < 5.25$ ppm. Although not shown, O_2 and N_2 distributions mirror exactly that of CO_2: $[O_2]$ $= 20.44 - 0.203[CO_2]$ with $r^2 = 0.995$; $[N_2] = 77.35 - 0.756[CO_2]$ with $r^2 = 0.995$. This indicates progressive dilution of essentially atmospheric air in the soil pores by the deep CO_2.

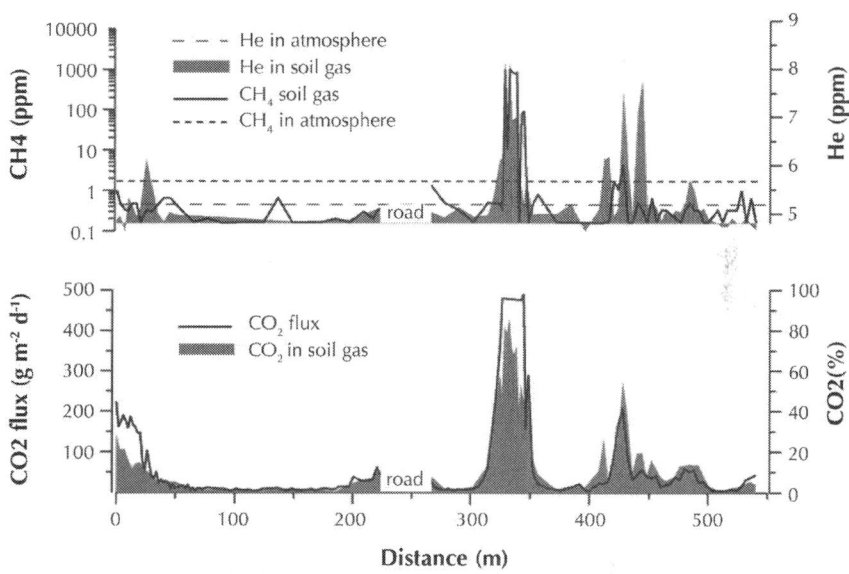

Figure. 7: Site 5 gas geochemistry profile conducted across a number of gas vents in the centre of the Latera caldera (see Fig. 2 for location). For reference the atmospheric concentration of He and CH_4 are represented as horizontal lines. Note that the scale for CO_2 flux has been truncated to show details at the lower range; as such six samples having flux values between 1000 and 3000 g m^{-2} d^{-1} have been removed in the interval from 330 to 340 m.

Detailed Study

The excellent exposure of a fault in the quarry wall of site 3 (Fig. 2) has allowed for highly detailed structural and CO_2 flux measurements, with the goal of understanding how small-scale structural features and fault evolution can control gas movement.

Structural Geology

The site 3 quarry is located in what appears to be volcanic lava flows (based on primary textures) that have been almost completely altered to kaolinite. The quarry face extends for about 100 m and has two main directions: N20°W and N70°E. As in the other quarries the deformation is essentially brittle, consisting of small faults and fractures that cluster around the two main directions illustrated above (N10°E and N50°E;Fig. 4); where observable faults have mainly dip–slip kinematics with a generally normal sense of shear.

The distribution of structural elements along the face of the quarry is variable. In general, fractures and faults are concentrated in 5–6 m wide domains having high deformation density, separated by intermediate domains where the density is lower. These high density fracture domains can be subdivided into three main configurations, which we believe represent different stages of the normal faulting process and/or the distribution of deformation along strike (i.e. evolution in time and/or space). As discussed below, these different styles have a significant influence on the gas migration pathways.

The first style is referred to as "single fault plane" deformation, and it is usually located within low-density fracture domains where primary structures are still well recognisable (Fig. 8a). It is characterised by a main slip plane that is polished and striated, and which sometimes has associated abrasive slickensides. The striations and orientations of the slickensides, where present, are consistent with a predominantly normal slip, with minor left- or right-lateral slip components depending on the local orientations of the surfaces. In this case the fault core (sensu Caine et al., 1996 and Chester et al., 1993) is very thin and poorly developed, a few millimetres thick, and composed of fine-grained material. The deformed rocks around the fault plane, i.e. the damage zone, consist of a 50–100 cm wide fracture zone. Fracture density around the fault plane is usually distributed asymmetrically, increasing towards the fault plane in the hanging wall compared to a relatively undeformed footwall. Deformation appears to be controlled exclusively by brittle processes and no evidence of fluid circulation on the slip plane or in the damage zone is recognisable.

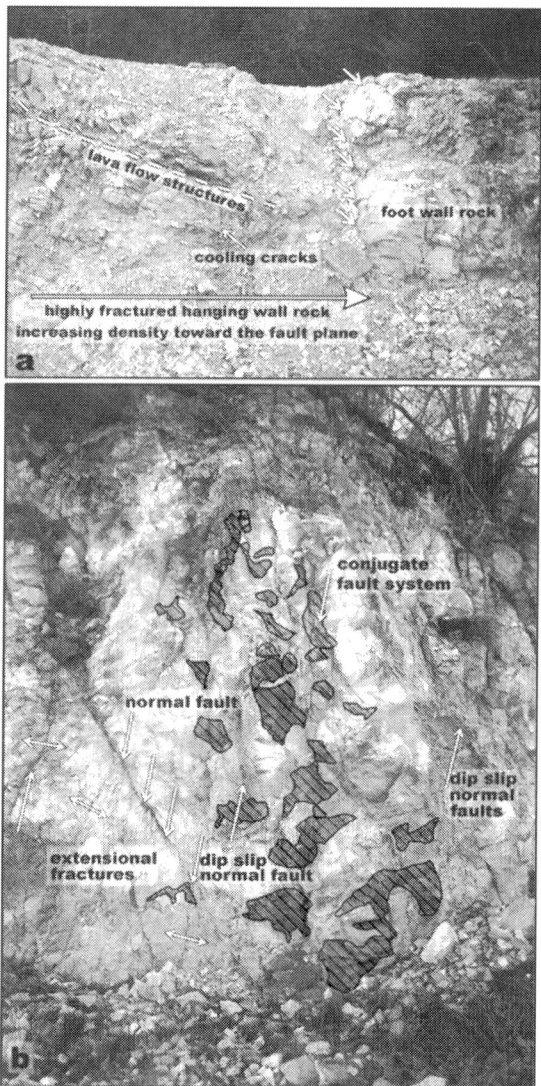

Figure. 8: (a) Photo of a single fault plane, illustrating the occurrence of the main, polished and striated slip plane (series of arrows) and associated hanging wall deformation; (b) photo of a fault network, with the structural assemblage composed of normal faults and a shear fracture

conjugate system. The hatch pattern highlights fault and fracture surfaces that are encrusted with sulphur precipitates.

The second style is referred to as "fault network" deformation, and it consists of an assemblage of normal faults and conjugate shear fracture systems having approximately the same strike (Fig. 8b). Fault plane kinematic indicators are mostly dip slip, and planes are generally at high angles to sigma 1. The fault network domains are 4–5 m wide and repeat every 8–10 m along the quarry face. Inside each domain the cross-cutting relationships among the fractures and striated slip surfaces are not univocal, suggesting that all of these structures belong to the same deformation event. Such synthetic and antithetic faults and fractures form a mesh (sensu Sibson, 1996 and Sibson, 2000) where interconnecting planes provide a conduit structure for fluid circulation. Although the preferred orientation for fluid circulation in meshes is parallel to the intersection of faults and fractures, as suggested by Sibson (1996), in this area fluid migrates to the surface from bottom to top. This is shown by the precipitation of sulphur crystals on the fault plane due to the upward migration of sulphur-rich fluids (hatched areas, Fig. 8b). Locally, where the network is not well developed and fractures are not completely interconnected, traces of sulphur crystals are observed on horizontal planes (parallel to fractures intersection) until a new vertical conduit is intersected.

The third deformation style is found in the central part of the quarry face, where a fault and its associated structures crops out (Fig. 9). This fault zone is almost symmetric and consists of two main slip planes which isolate the 5 m wide fault core from the surrounding 5–6 m wide damage zones. The two walls of the core are quite regular, planar-slip surfaces that are approximately parallel, sub-vertical, dip towards 150°, and have abrasive slickensides that indicate a normal sense of shear. The core zone comprises materials which vary in grain size with an almost symmetric distribution, consisting of mainly grey, clayey material mixed with small lithic fragments (about 10–20 mm in diameter), as well as clays, white pumice, and other black volcanic rocks. In the central part, up to 1 m diameter blocks of the host rock are encompassed in the clayey material, as are blocks of clays and other black lavas. The following sequence can be recognised from one slip plane to the other: 2 m of fine material, 1.5 m of breccias (including blocks of the host rock), and again about 1 m of fine material.

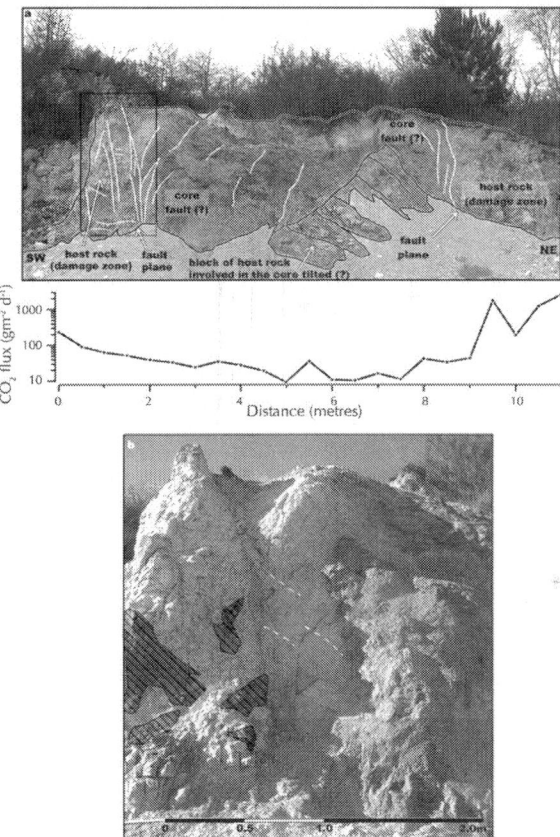

Figure. 9: Mature fault in the site 3 quarry and the distribution of CO_2 flux across it (a). Note the symmetry of the fault with two damage zones and a central part consisting of clayey material. The two main faults, the blocks of the host rocks involved inside the core zone, and the deformation close to the wall faults are highlighted. (b) Detail of one of the faults outlined in (a), showing the bands inside the clayey material and their drag geometry close to the slip planes as well as the small and poorly developed slip surfaces. The fault/fracture surfaces stained by sulphur are

outlined. On both sides of the core the clayey material is non-cohesive. Here lithic fragments are observed as aligned bands of different grain sizes that dip towards 300°, showing a trend that is parallel to the fault wall. Close to the slip planes these bands exhibit a drag geometry that is compatible with the kinematics observed on the slip planes themselves. Small, smooth, and poorly developed slip surfaces present inside this zone cross-cut each other (Fig. 9b) and cut the mentioned lithic bands. These slip surfaces have the same attitude as the main slip planes, show abrasive kinematic indicators with normal, dip–slip movement, and could represent zones of higher cohesion and localised shear stress; these features do not form a network inside the core zone. In general, fracture density in the clayey material is very low, and planes are often filled by reddish to black material that may also be related to surface water infiltration. In the central part, greater fracturing observed in host rock blocks does not continue into the clayey material.

Fracturing in the damage zone around the fault core is pervasive, and planes are clearly connected to each other as described for the fault network domain above (Fig. 8 and Fig. 9). This interval consists of a conjugate system of high-angle fractures and normal faults parallel to the main slip planes. These parts of the fault represent the main fluid conduit, as shown by the precipitation of sulphur minerals on the planes.

The large contrast between the lithic, cohesive host rock and the clayey material in the core might be explained not only by mechanical re-working and geochemical reactions induced by water and gas movement, but also by the simple filling of a soft, fault-core depression with fine-grained material. However, both the clayey material and the damage zones are deformed, and so even if some of the core material was originally deposited in this interval the final composition and structural configuration is to be attributed to mechanical and geochemical processes. Mineralogical and geochemical analyses of this composite material are in progress to better define these processes.

CO₂ Flux

In total, 267 CO_2 flux measurements were collected along a regular grid on the quarry floor in front of the structures described above, with the results summarised in Table 1. The mean CO_2 flux value (1700 g m⁻² d⁻¹) is very high due to the maximum value of 49563 g m⁻² d⁻¹, while the median (331 g m⁻² d⁻¹) is also very high because almost the entire gridded area of the quarry floor has anomalous flux rates. Data analysis with normal probability plots shows that a total of five different populations can be defined, each of which is linked to a particular migration environment. Only about 10% of the samples are in the background population (defined as 0–60 g m⁻² d⁻¹), whereas all others can be considered as anomalous.

The extreme spatial variability of the site is highlighted by flux rate values that change by 2 orders of magnitude between sampling points separated by 3 m, while a separate test showed a change of more than 1 order of magnitude over only 30 cm. This clearly illustrates the "spot" nature of gas migration along spatially restricted channels (or "pipes") along the fault plane.

Directional experimental variograms conducted on the log-transformed CO_2 flux data highlight an anisotropy having a N50°E maximum axis. The computed experimental variograms were modelled along the maximum anisotropy axis, yielding the following equation: $\gamma = 0.32 + 0.2$ Sph (42 m) and an anisotropy ratio of 2. These results were used as parameters for the kriging calculations, which produced the CO_2 flux contour map given in Fig. 10. This map clearly illustrates a N50°E trend, while the shape of the anomalies confirms the highly anisotropic behaviour.

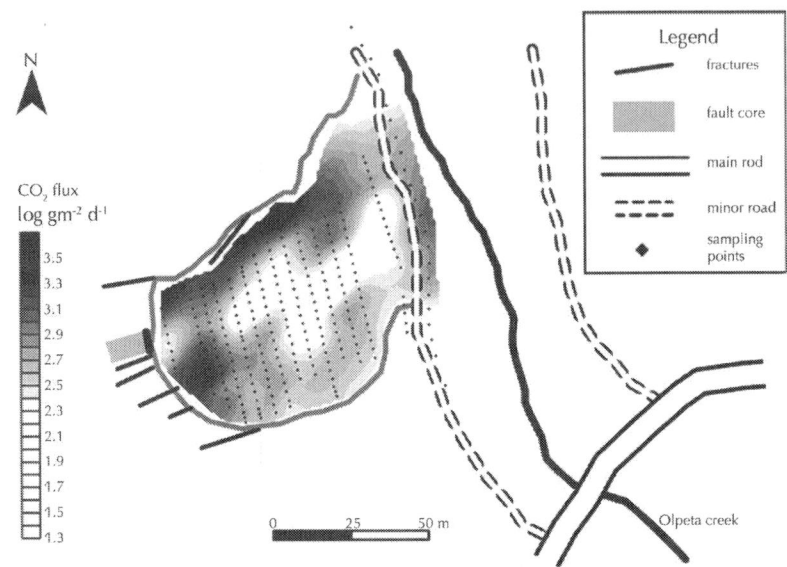

Figure. 10: The contoured distribution of CO_2 flux in the site 3 quarry. Note the mapped orientation of fractures on the SW face of the quarry, including the interval of the "mature fault" shown by the grey box. Black dots show the sampling locations.

DISCUSSION

The Structural Control of Gas Migration

The structural work performed throughout the Latera caldera area has defined normal kinematics along two principle trends, one at N10°E and the other at N50°E (Fig. 4), which agree with the regional trends observed throughout northern Latium and in deep boreholes drilled in the caldera area itself for geothermal exploration (Sabatelli and Mannari, 1995). At the surface these features tend to be open in the brittle volcanic units in which they are found, providing a highly permeable pathway for CO_2 in the geothermal reservoir to escape towards the atmosphere.

This is clearly illustrated by the fact that the regional soil gas results (Fig. 6) also show preferential alignment between N–S and N50°E. In this figure, the values that are greater than the 2.5% contour can be considered as anomalous, based on an analysis of the normal probability plot for this dataset. Values below this threshold concentration are likely due to soil biological processes such as microbial and root respiration (e.g. Risk et al., 2002).

Although this figure, together with the geostatistical analysis, shows lineament directions that are similar to those of the measured faults and fracture systems, leakage does not occur along the entire length of any given fault but is instead focused at isolated points or small areas known as gas vents. This implies a complex permeability distribution along the fault planes, whereby gas flow occurs via preferential pathways consisting of well-interconnected, larger voids (i.e. gas is the non-wetting phase). The existence and location of these pathways, and the eventual gas release points on the surface, is controlled by a number of processes. At the pore scale these would include permeability reduction via secondary mineral precipitation (Cavarretta et al., 1985) or clay infilling (this study). At a larger scale, issues related to fault intersection and fault development can influence the distribution and nature of secondary permeability, as discussed below.

Fault intersection can result from one fault cross-cutting an older, unrelated fault that has a different orientation, or via the contact between a main fault and its related transfer structure. In either case this juncture could have increased dilation or brecciation, creating a more permeable, vertical, gas migration pathway above which soil gas anomalies may form (e.g. Fountain and Jacobi, 2000). The fault and fracture orientations measured in the Latera caldera cluster around the two directions of N10°E and N50°E, indicating the potential for intersecting structures. In addition, cross-cutting conjugate fracture patterns are observed in outcrop, and small-scale vertical offset is locally transferred from one main fault to another via high angle, normal/strike-slip faults. Based on observations of rock exposures, the N50°E fractures and faults appear to be relay ramps associated with the main fault system at N10°E. This seems to be in agreement with the variogram analysis of the regional soil gas CO_2 data, where the small-scale main anisotropy axis trends N45°E and the large-scale main anisotropy axis trends N20°E.

If gas migration pathways occur along the intersection of two faults, one might expect them to be more "pipe-like" (along the intersection) and to release gas at the surface from more spatially isolated spots. One gas vent that may meet these criteria is the so-called "gas vent A" (GVA), which has been studied extensively in Beaubien et al. (2008) and which occurs between 300 and 350 m in the profile shown in Fig. 7. This vent is clearly isolated, with a small (<10 m) core of high CO_2 flux rates (1000–3000 g m^{-2} d^{-1}) surrounded by a transition zone in which flux values return rapidly to background levels. A detailed soil gas grid conducted above this vent (data not shown) shows two anomaly orientations, implying that they may have formed at the juncture of two structures.

Fault development, along strike or in time, can also greatly influence secondary permeability due to changes in fracture interconnectivity and self-sealing processes. Such variations can be seen in the site 3 quarry, where the three-dimensional exposure allows for a better understanding of structure distribution and its influence on gas migration pathways. As described, three different deformation styles are observed here: style 1—simple, spatially restricted, single fault planes; style 2—more complex fault networks, where fractures are more developed, inter-connected, and secondary permeability is high; and style 3—mature faults, where the system has evolved a clay-rich fault core that is relatively impermeable.

Highly detailed gas flux measurements on the quarry floor show a distribution that agrees well with the structural observations made on the quarry walls (Fig. 4). In the middle of the quarry floor an interval of low CO_2 flux extends from the central, clay-rich mature fault core (i.e. style 3) in a direction that is essentially the same as the measured fracture patterns. On the northern edge of this low-flux interval there is an alignment (N50°E) of extremely high values (up to 40,000 g m^{-2} d^{-1}) that lie along strike with the damage zone of the main fault (which is similar to the style 2 deformation); along the quarry face this interval shows highly brittle deformation with large cavities and open, interconnected fractures. Finally, some isolated anomalies appear to correspond with style 1 deformation, however these tend to be less continuous and less pronounced.

The effect of these different styles can be seen schematically in Fig. 11. Here the distribution of CO_2 soil gas or flux above a fault is shown to be controlled by the state of fault evolution (e.g. styles 2 or 3). If style 2 dominates there will be a well-developed fracture network on which gas migration will tend to be centred (Fig. 11a). Where such a feature is buried beneath shallow sediments, a distribution may occur like that observed from 400 to 500 m along the site 5 profile (Fig. 7). Here the anomaly is wide, irregular, and values are generally much lower than those observed in isolated GVA, implying a wide fracture network where gas flux maximums are reduced due to the division of flow into a larger number of escape pathways.

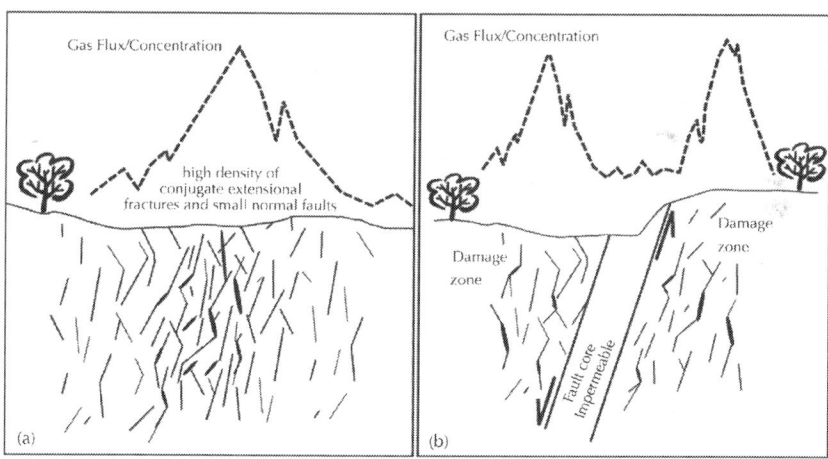

Figure. 11: Schematic drawing showing different structural assemblages and their influence on CO_2 leakage; these stages of deformation could represent changes either in time as the fault develops, or along strike due to differential movements. (a) The fault network configuration results in more open, interconnected gas migration pathways, resulting in higher gas flux rates. (b) The mature fault shows the development of a very low

permeability core, bounded by damage zones that maintain the fault network character.

Instead, if the feature is a mature fault with a well developed cataclastic/gouge core bounded by a highly fractured damage zone (i.e. style 3), values above the low-permeability fault core would be low and two peaks would occur above the lateral fracture zones (Fig. 11b), as is seen in the site 3 quarry (Figs. 9a and10). Such a double peak distribution has also been observed in profiles conducted across the San Andreas Fault in California, with a similar fault architecture being proposed as the controlling mechanism (King et al., 1996). Although recent research has shown that these observed anomalies may not be due to deep gas migration, but rather to wind-induced movement of biogenic and near-surface gases within the bounding fracture zones (Lewicki and Brantley, 2000 and Lewicki et al., 2003), the structural architecture still remains valid.

It should be remembered that the three styles observed at site 3 may represent not only an evolution in time as the fault matures, but also in space due to differential movements or the intersection of different lithologies (e.g. Baubron et al., 2002). In terms of spatial distribution, this along-strike change in permeability would have a fundamental impact on the nature of surface CO_2 release, with gas flux or soil gas anomalies being perhaps aligned but not continuous, or parallel but not aligned (e.g. Ciotoli et al., 2007). In between these anomalies there may be no gas flow, although individual "spot anomalies" may begin to coalesce if they are sufficiently close to each other (e.g. Lewicki et al., 2007c). An example of a no-flow regime is shown by the long interval of background soil gas concentrations (CO_2 < 2.5%, He < 5.25 ppm; CH_4 < 1.5 ppm) and CO_2 flux values (<22 g m^{-2} d^{-1}) observed from 50 to 200 m in Fig. 7, indicating that there are large blocks of intact volcanics (and overlying sediments) that impede deep CO_2 migration.

Thus the results given in Fig. 7 indicate that gas release at surface is primarily controlled by faults, and argue against large-scale diffuse degassing (sensu Chiodini et al., 2007) in this area (although more research is needed to better understand if it is occurring elsewhere). Instead, the size of an anomaly appears to be controlled by lateral migration and chemical/biological processes occurring in the vadose zone, as discussed below.

Gas Migration through the Vadose Zone

Conceptually, gas ascending from the geothermal reservoir at Latera will migrate as bubble or continuous gas flow along faults and fracture systems until the water table is encountered, after which migration will occur via gas-phase advective or diffusive flow in the unsaturated vadose zone. Gas will also dissolve into the groundwater and then be transported laterally via groundwater flow, however this mechanism is not addressed here. The vadose zone can either consist of bedrock, such as that observed in the site 3 quarry, or of shallow sediments and soil, as occurs at site 5. Depending on the thickness, porosity and permeability, water content, microbiology, and chemistry of the unsaturated zone, the deep-gas signature originating from the fault will be altered to some degree. Such a change will influence our ability to locate CO_2 leakage (especially from buried faults) with any near-surface monitoring technique, such as gas-geochemistry, atmospheric, biological, or remote-sensing measurements. The following discusses the influence of the vadose zone at the Latera site.

The distribution of anomalous soil gas CO_2 is wider than the intervals of highest CO_2 flux, as illustrated by a detailed portion of the site 5 profile (Fig. 12). The wide interval of anomalous soil gas CO_2 (from about 300 to 370 m), as compared to the core of very elevated CO_2 flux values (from 328 to 342 m), is likely linked to the horizontal migration of soil gas CO_2 via both advective forces and density-driven flow. For example a numerical modelling study conducted by Altevogt and Celia (2004) found that vertical and horizontal pressure gradients become similar at higher flow rates and that the density contrast between air and CO_2 can result in more lateral spreading and storage of CO_2 in the unsaturated zone.

Similar lateral distributions in the vadose zone were also found by Lewicki et al. (2007c) in simulations of a CO_2 release slightly below the water table. The observed distribution of CO_2 may also be slightly attenuated from even-further lateral migration due to mass loss via dissolution into pore water and subsequent involvement in acidic reactions with the sediments (Altevogt and Jaffe, 2005).

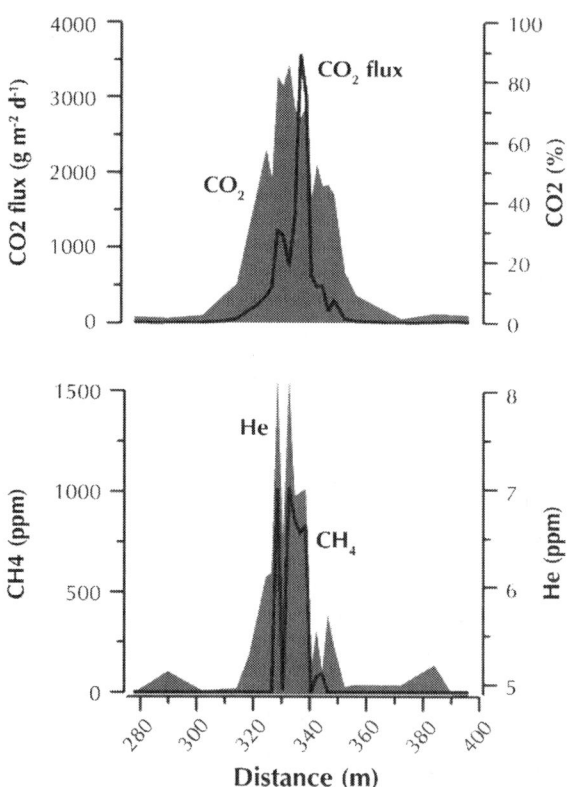

Figure. 12: Detail of the site 5 profile. Compared to the complete profile given in Fig. 7, this figure shows the CH_4 data plotted on a normal scale and the CO_2 flux data are not truncated.

Other proposed mechanisms for increased soil gas CO_2 values peripheral to a vent include the return of released CO_2 via dissolution in falling rainwater or back diffusion from the atmosphere (Oldenburg and Unger, 2004). As sampling was conducted during the dry season, infiltration need not be considered here, whereas the steady increase in CO_2 concentrations with depth in vertical profiles conducted around the vent (not shown) argues against a major contribution from back diffusion from the atmosphere.

Fig. 12 shows the distribution of anomalous He values to be narrower than that of CO_2, but in good correspondence with the interval of moderately elevated CO_2 flux values (i.e. from 312 to 355 m). The slightly lower lateral migration of He, compared to CO_2, may be the result of such factors as: (i) helium has a lower density than air, and thus density-driven flow is not an issue but instead buoyancy forces may result in an additional vertical flow component; (ii) helium is much less soluble than CO_2, and thus there is no potential for storage in pore water and subsequent equilibrium exchange with the gas phase; and (iii) the ratio between deep gas and background soil air concentration is much lower for He (<2) than for CO_2 (about 100) at this site, and thus He diffusion gradients will be lower (thereby minimising the effect of the higher He diffusion coefficient). It should also be remembered that He is non-reactive, and thus no chemical attenuation mechanism exists like that described above for CO_2. Such considerations may also be important for the use of other tracer gases that have been proposed to monitor for leakage from CO_2 geological storage sites, such as sulphurhexafluoride (SF_6) or various perfluorocarbon gases (e.g. Wells et al., 2007), as their physical–chemical characteristics are similar to those of He (i.e. non-reactive, highly mobile, etc.). That said, artificial tracers have the advantage of very low natural concentrations, thus detection limits are very low and there is the potential for large concentration gradients that may promote lateral diffusion.

The distribution of anomalous CH_4 concentrations is even more restricted than that of He, as shown by the limited number of samples along the site 5 profile (Fig. 7) with values above that of atmospheric air (1.8 ppm): at 425 m (4.4 ppm) and in gas vent A from 325 to 345 m (up to 1000 ppm). The CH_4 results for GVA are more clearly seen in Fig. 12, where they are plotted on a normal scale. Here one can see that anomalous CH_4 values exist at 80 cm depth only in the interval where CO_2 flux values are at their maximum (from 328 to 342 m). In contrast to both CO_2 and He, CH_4 is redox reactive and readily oxidised in aerobic soils (e.g. Castaldi et al., 2007), meaning that CH_4 concentrations in well-aerated soils are typically below atmospheric concentrations. As discussed in more detail by Beaubien et al. (2008), this reactive gas (and others such as hydrogen sulphide, hydrogen, and ethane) is maintained in the anoxic core of the vent, but is rapidly oxidised as it migrates laterally into zones with higher O_2 concentrations. Considering its distribution in Fig. 7 and Fig. 12, it appears that a CO_2 flux rate greater than 500 g m^{-2} d^{-1} is needed to create an environment, in these soils and at this depth, in which CH_4 remains stable.

Although the analysis of as many gas species as possible will improve the potential for identifying and locating a leak above a geological CO_2 storage site, the effectiveness of an individual gas species may be influenced by its spatial distribution and the size of anomaly that it may produce (as described above). In addition, any monitoring program above a CO_2 geological storage site must try to enhance the potential for success while at the same time keeping costs acceptable. To balance these two issues of success rate and costs, and considering the observation that CO_2 has the largest "footprint" around a leak while other gas species are more spatially restricted (but potentially more indicative of a deep origin), one could consider the following approach: a large number of points could be measured in the field for CO_2 flux and soil gas CO_2, and then at locations where field results are above a pre-set threshold a soil gas sample could also be collected for the laboratory analysis of other species (and/or isotopes). As an experienced technician can do field measurements at 30–70 points per day (depending on terrain and sample spacing) there is the potential to measure a large number of sites with this approach, which will allow for a higher sample density and a greater potential for locating what may be a spatially small anomaly. Subsequently, the laboratory analysis of various tracer gases (He, CH_4, artificial tracers, isotopes, etc.) on only those samples having anomalous CO_2 flux or concentration will then help define the origin of the anomaly (deep or shallow) while at the same time keeping costs low. Subsequent detailed sampling grids or profiles could also be measured around an anomaly to help confirm its origin and to define its lateral extent. The types of gases used for separating deep – from shallow – origin CO_2 anomalies should be chosen based on the type of storage reservoir being studied. For example, CH_4 and He would be good choices for monitoring above a CO_2-EOR site because they are naturally associated with the reservoir, whereas artificial tracers added to the CO_2 injection stream might be preferred above a deep saline aquifer.

It must be stressed, however, that this approach is based on the results obtained from an essentially steady-state system having site-specific conditions (e.g. vadose zone thickness, sediment permeability and composition, climate and rainfall, gas flow rate and composition, etc.); it is recommended that similar work be conducted on other sites with different settings to extend the results obtained here.

CONCLUSIONS

Structural geology and near-surface gas geochemistry surveys conducted within the Latera caldera, where naturally produced CO_2 is leaking to the atmosphere, have highlighted a number of issues related to the migration of gas along faults and its eventual transfer to the atmosphere.

Both detailed and regional soil gas and gas flux measurements have the same general trends as those measured in outcrops and in deep boreholes (approximately N10°E and N45°E), showing that migration is controlled by these structures. Release at surface, however, is not continuous along strike, but instead occurs at determined places due to channelled flow along more permeable pathways along the fault plane. Work on an exposed fault shows a clear link between deformation style, secondary permeability characteristics, and the spatial distribution and mass transfer rates of deep gas to the surface. These styles represent not only temporal changes in a fault system during its evolution, but may also be the result of spatial changes caused by differential movements or the intersection of different lithologies along strike; this variability clearly has an influence on the distribution of gas migration at surface. For example a detailed gas geochemistry profile conducted across buried faults showed different types of venting features, ranging from isolated, high-flux-rate gas vents that may have formed at the juncture of two faults (concentrated "pipe" flow) to wider areas of more diffuse degassing which may be related to a wide damage zone of networked fractures (diffuse "sheet" flow). Wide intervals occur between these degassing sites where all measured parameters are at background values, indicating that there are unfaulted blocks of volcanic rocks and overlying sediments that represent barriers to upward gas migration.

Movement of the deep gases in the vadose zone was also examined, particularly with detailed sampling across an isolated gas vent (GVA). These results show how the CO_2 flux rate and distribution is important for controlling the subsurface distribution of the various migrating gases, but that the individual chemical–physical characteristics of each gas results in different behaviours. Soil gas CO_2 at 80 cm depth was found to have the widest lateral distribution (mirrored by opposite O_2 and N_2 trends), He was narrower, while reactive gas species like CH_4 and H_2S were very narrow and restricted to the highest CO_2 flux interval in the vent core. Different diffusion coefficients, redox reactivity, gas density, and concentration gradients are believed to be some of the more important controlling parameters.

The results of this study have many implications for CO_2 geological storage. To begin with the necessity of a thorough knowledge and understanding of the structural geology of any site considered for geological CO_2 storage has been highlighted through the close link found between faults and gas migration in the Latera caldera. Such site characterisation, which would help to focus future monitoring on areas which might have a higher risk, could also include near-surface gas geochemistry techniques to search for any gas-permeable structures. As seen here regional surveys have the potential to outline anomalies, depending on the sampling density and size of any eventual leak. However any results must be subsequently checked using detailed sampling grids or horizontal profiles. In addition, the distribution of leaking deep gases in the soil has been shown to depend both on vadose zone characteristics and the physical–chemical properties of each gas species, and that this information should be considered in the design of any near-surface gas geochemistry monitoring program.

This research has shown how the study of natural systems like that at Latera permits us to better understand gas migration relative to the issues of scale, heterogeneity, and system evolution, thus improving our ability to choose, construct, and monitor safe CO_2 geological storage sites. Even if a volcanic terrain will likely never be chosen for a geological storage site, it is believed that the issues raised in this article and many of the processes and mechanisms observed at Latera can be used in other geological settings.

ACKNOWLEDGEMENTS

We would like to thank Anna Baccani for her help during the laboratory analyses, Mr. Iacarelli for his kind access to one of the study sites, and Lucio Poscia of the local Latera administration for his support in the field and for giving access to the abandoned quarry at site 3. Some data presented here was collected within the European-Community funded project on natural analogues of CO_2 geological storage (NASCENT), while the rest was conducted within the European-Community funded Network of Excellence on CO_2 geological storage (CO2GeoNet). The authors would also like to sincerely thank two anonymous reviewers for their thoughtful comments and suggestions, which significantly improved the manuscript.

APPENDIX A. GEOLOGICAL SETTING OF THE LATERA CALDERA

A.1. Pre-Volcanic Sedimentary Sequences

There are five main sedimentary sequences that were deposited in the area prior to volcanic activity: (i) Triassic to Oligocene marine sediments of the Tuscan domain, consisting of Triassic to Lower Cretaceous limestones and marls (Tuscan sequence) and Cretaceous to Oligocene sandstones, turbiditic limestones and argillites (Tolfa Flysch); (ii) Triassic to Oligocene marine sediments of the Umbria domain, consisting of evaporites, limestones and marls; (iii) Lower Cretaceous to Paleocene allochthonous flysch of the Ligurid domain, consisting of marine shales, siltstones and marls with subordinate sandstones; (iv) Miocene continental sediments consisting of conglomerates and shales deposited prior to volcanic activity; and (v) Upper Pliocene–Upper Calabrian marine to brackish sediments, represented by clays and sandy-clays with polygenic conglomerates (Bertini et al., 1971). These sedimentary formations randomly crop out in the Vulsini Volcanic district, although they are more continuous near Mt. Razzano, where a NS-syncline deforms the Tuscan formations (Buonasorte et al., 1991), and Mt. Canino, where the Liassic to Cretaceous Tuscan sequence forms a NS-trending anticline and syncline. The same lithologies are crossed in deep geothermal exploration wells throughout the area.

A.2. Volcanic Sequences

The Latera Volcanic complex developed in the western part of the Vulsini district between 0.3 and 0.1 Ma (Palladino and Simei, 2005, Simei et al., 2006 and Vezzoli et al., 1987). Although the Latera caldera has a relatively simple topography, its eruptive history is complex (Fig. 3), with multiple depocentres and a close link to Quaternary extensional tectonics. The main steps of the caldera evolution are related to three main eruptions (Sovana, Onano and Pitigliano; Fig. 3) that occurred from 0.19 to 0.16 Ma ago (Palladino and Simei, 2005). The Latera caldera is the main edifice to the west of the Bolsena caldera, and it marks a change from previously dominant effusive volcanism (around 0.4–0.3 Ma) to explosive volcanism characterised by a wide spectrum of potassic rock types. This explosive nature resulted in a polygenic collapse caldera which has been dated at 0.3 Ma (Palladino and Agosta, 1997, Palladino and Simei, 2005,Palladino and Valentine, 1995, Sparks, 1975, Trigila and Walker, 1981 and Vezzoli et al., 1987), followed by recurrent Plinian activity 0.28–0.23 Ma which formed a central volcanic edifice Palladino and Agosta, 1997 and Palladino and Simei, 2005). The subsequent Sovana eruption resulted in pyroclastic flow activity (around 0.19 Ma) and is related to the main collapse of the Latera caldera. The largest volume of volcanic products was then erupted during the Onano eruption (0.17 Ma), which occurred from a fissure-like vent bordering the Latera caldera to the east instead of a central volcanic edifice (Freda et al., 1990, Nappi, 1969 and Vezzoli et al., 1987). The final phases of activity were characterised by the migration of emission epicentre from east to the northwest, where the Pitigliano eruption took place at 0.16 Ma and generated the small Vepe caldera (Nappi et al., 1991). Subsequent activity was then limited to effusive and hydromagmatic eruptions from multiple monogenetic centres both within and outside of the Latera caldera (Palladino and Simei, 2005).

REFERENCES

1.   Agosta, F., Aydin, A., 2006. Architecture and deformation mechanism of a basin-bounding normal fault in Mesozoic platform carbonates, central Italy. Journal of Structural Geology 28, 1445–1467.

2.   Altevogt, A.S., Celia, M.A., 2004. Numerical modelling of carbon dioxide in unsaturated soils due to deep subsurface

leakage. Water Resources Research 40 (W03509), doi:10.1029/ 2003WR002848.

3.   Altevogt, A.S., Jaffe, P.R., 2005. Modelling the effects of gas phase CO2 intrusion on the biogeochemistry of variably saturated soils. Water Resources Research 41 (W09426), doi:10.1029/ 2004WR003819.

4.   Anderson, D.E., Farrar, C.D., 2001. Eddy covariance measurement of CO2 flux to the atmosphere from an area of high volcanogenic emissions, Mammoth Mountain, California. Chemical Geology 177, 31–42.

5.   Antonellini, M., Aydin, A., Pollard, D., 1994. Microstructure of deformation bands in porous sandstones at Arches National Park, Utah. Journal of Structural Geology 16, 941–959.

6.   Azzaro, R., Branca, S., Giammanco, S., Gurrieri, S., Rasa, R., Valenza, M., 1998. New evidence for the form and extent of the Pernicana Fault System (Mt. Etna) from structural and soil-gas surveying. Journal of Volcanology and Geochemical Research 84, 143–152.

7.   Baines, S.J., Worden, R.H., 2004. The long-term fate of CO2 in the subsurface: natural analogues for CO2 storage. In: Baines, S.J., Worden, R.H. (Eds.), Geological Storage of Carbon Dioxide, Special Publication, 233. The Geological Society of London, London, pp. 59–86.

8.   Barberi, F., Innocenti, F., Landi, P., Rossi, U., Saitta, M., Santacroce, R., Villa, I.M., 1984. The evolution of Latera caldera (central Italy) in the light of subsurface data. Bulletin of Volcanology 47, 125–141

9.   Barton, C.A., Zoback, M.A., Moos, D., 1995. Fluid flow along potentially active faults in crystalline rock. Geology 23, 683–686.

10.   Bateson, L., Vellico, M., Beaubien, S.E., Pearce, J.M., Ciotoli, G., Annunziatellis, A., Coren, F., Lombardi, S., Marsh, S., 2008. Preliminary results of the application of remote sensing techniques to detecting and monitoring leaks from CO2 storage sites. International Journal of Greenhouse Gas Control

2 (3), 388–400.

11. Baubron, J.C., Rigo, A., Toutain, J.P., 2002. Soil gas profiles as a tool to characterise active tectonic areas: the Jaut Pass example (Pyrenees, France). Earth and Planetary Science Letters 196, 69–81.

12. Beaubien, S.E., Ciotoli, G., Coombs, P., Dictor, M.C., Krü̈ger, M., Lombardi, S., Pearce, J.M., West, J.M., 2008. The impact of a naturally occurring CO2 gas vent on the shallow

13. ecosystem and soil chemistry of a Mediterranean pasture (Latera Italy). International Journal Greenhouse Gas Control 2 (3), 373–387.

14. Beaubien, S.E., Ciotoli, G., Lombardi, S., 2003. Carbon dioxide and radon gas hazard in the Alban Hills area (central Italy). Journal of Volcanology and Geothermal Research 123, 63–80.

15. Beaubien, S.E., Lombardi, S., Ciotoli, G., Annunziatellis, A., Hatziyannis, G., Metaxas, A., Pearce, J.M., 2005. Potential hazards of CO2 leakage in storage systems—learning from natural systems. In: Rubin, E.S., Keith, D.W., Gilboy, C.F. (Eds.), Proceedings of the 7th International Conference on Greenhouse Gas Control Technologies, vol. 1. PeerReviewed Papers and Plenary Presentations. Elsevier, Oxford, UK, pp. 551–560.

16. Bergfeld, D., Evans, W.C., Howle, J.F., Farrar, C.D., 2006. Carbon dioxide emissions from vegetation-kill zones around the resurgent dome of Long Valley caldera, eastern California,

17. USA. Journal of Volcanology and Geothermal Research 152, 140–156, doi:10.1016/j.jvolgeores.2005.11.003.

18. Bertini, G., D'Amico, C., Deriu, M., Tagliavini, S., Vernia, L., 1971. Note illustrative della Carta Geologica d'Italia alla scala 1:100.000. Foglio 143 Bracciano. Roma, Serv. Geol. d'Italia.

19. Bertrami, R., Cameli, G.H., Lovari, F., Rossi, U., 1984. Discovery of Latera geothermal field: problems of the exploration and research. In: Seminar on Utilization of

Geothermal Energy for Electric Power Production and Space Heating, Florence, 14–17 May 1984.

20. Brennan, S.T., Hughes, A.V., Friedmann, S.J., Burruss, R.C., 2004. Natural gas reservoirs with high CO_2 concentrations as natural analogues for CO_2 storage. In: The 7th Greenhouse Gas Control Technologies Conference (GHGT7), Vancouver, Canada, 5–9 September 2004.

21. Bruhn, R.L., Parry, W.T., Yonkee, W.A., Thompson, T., 1994. Fracturing and hydrothermal alteration in normal fault zones. Pure and Applied Geophysics 142, 611–643.

22. Buonasorte, G., Ciccacci, S., De Rita, D., Fredi, P., Lupia Palmieri, E., 1991. Some relations between morphological characteristics and geological structure in the Vulsini

23. Volcanic Complex (Northern Latium, Italy). Zeitschrift fuer Geomorphologie Supplement BD 82, 59–71.

24. Caine, J.S., Evans, J.P., Forster, C.B., 1996. Fault zone architecture and permeability structure. Geology 24, 1025–1028.

25. Caine, J.S., Forster, C.B., 1999. Fault zone architecture and fluid flow: insight from field data and numerical modeling. In: Haneberg, W.C., Mozley, P.S., Moore, J.C., Goodwin, L.B.

26. (Eds.), Faults and Subsurface Fluid Flow, Geophysical Monograph 113. pp. 101–128.

27. Carmignani, L., Kligfield, R., 1990. Crustal extension in the Northern Apennines: the transition from compression to extension in the Alpi Apuane core complex. Tectonics 9, 340–349.

28. Castaldi, S., Costantini, M., Cenciarelli, P., Ciccioli, P., Valentini, R., 2007. The methane sink associated to soils of natural and agricultural ecosystems in Italy. Chemosphere 66, 723–729.

29. Cavarretta, G., Giannelli, G., Scandiffio, G., Tecce, F., 1985. Evolution of the Latera Geothermal System II: metamorphic,

hydrothermal mineral assemblages and fluid chemistry. Journal of Volcanology and Geochemical Research 26, 337–364.

30. Chadwick, A., Arts, R., Eiken, O., Williamson, P., Williams, G., 2006. Geophysical monitoring of the CO2 plume at Sleipner, North Sea. In: Lombardi, S., Altunina, L.K., Beaubien, S.E. (Eds.), Advances in the Geological Storage of Carbon

31. Dioxide. NATO Science Series. IV. Earth and Environmental Sciences, vol. 65. Springer, Dordrecht, The Netherlands, pp. 303–314.

32. Chester, F.M., Evans, J.P., Biegel, R.L., 1993. Internal structure and weakening mechanisms of the San Andreas faults. Journal of Geophysical Research 98, 771–786.

33. Chester, F.M., Logan, J.M., 1986. Implications for mechanical properties of brittle faults from observations of the Punchbowl fault zone, California. Pure Applied Geophysics 24, 79–106.

34. Chiodini, G., Baldini, A., Barberi, F., Carapezza, M.L., Cardellini, C., Frondini, F., Granieri, D., Ranaldi, M., 2007. Carbon dioxide degassing at Latera caldera (Italy): evidence of

35. geothermal reservoir and evaluation of its potential energy. Journal of Geophysical Research 112 (B12204), doi:10.1029/2006JB004896.

36. Chiodini, G., Frondini, F., 2001. Carbon dioxide degassing from the Albani Hills volcanic region, central Italy. Chemical Geology 177, 67–83.

37. Chiodini, G., Frondini, F., Ponziani, F., 1995. Deep structures and carbon dioxide degassing in central Italy. Geothermics 24, 81–94.

38. Ciotoli, G., Lombardi, S., Annunziatellis, A., 2007. Geostatistical analysis of soil gas data in a high seismic intermontane basin: the Fucino Plain, central Italy. Journal of Geophysical Research 112 (B05407), doi:10.1029/2005JB004044.

39. Cowie, P.A., Scholz, C.H., 1992. Growth of faults by accumulation of seismic slip. Journal of Geophysical Research

97, 11085–11095.

40. Di Filippo, M., Lombardi, S., Nappi, G., Reimer, G.M., Renzulli, A., Toro, B., 1999. Volcano tectonic structures, gravity and helium in geothermal areas of Tuscany and Latium (Vulsini volcanic district), Italy. Geothermics 28, 377–393.

41. Duchi, V., Minissale, A., Paolieri, M., Prati, F., Valori, A., 1992. Chemical relationship between discharging fluids in the Siena-Radiocfani Graben and the deep fluids produced by the geothermal fields of Mt. Amiata, Torre Alfina and Latera. Geothermics 21 (3), 401–413.

42. Evans, J.P., Foster, C.B., Goddard, J.V., 1997. Permeability of faultrelated rocks, and applications for hydraulic structure of fault zones. Journal of Structural Geology 19, 1393–1404.

43. Evemden, J.F., Curtis, G.H., 1965. K/Ar of late Cenozoic rocks in Eastern Africa and Italy. Current Anthropology 6, 343–385.

44. Flodin, E.A., Gerdes, M., Aydin, A., Wiggins, W.D., 2005. Petrophysical properties of cataclastic fault rock in sandstone. In: Sorkhabi, R., Tsuji, Y. (Eds.), Faults, Fluid Flow, and Petroleum Traps. AAPG Memoir, vol. 85. pp. 197–227.

45. Fountain, J.C., Jacobi, R.D., 2000. Detection of buried faults and fractures using soil gas analysis. Environmental and Engineering Geoscience 6 (3), 201–208.

46. Freda, C., Palladino, D.M., Pignatti, S., Trigila, R., Onorati, G., Poscolieri, M., 1990. Volcano-tectonic scenario of Vulsini Volcanoes (Central Italy) from LANDSAT-MSS images and digital elevation data. Journal of Photogrammetry and Remote Sensing 45, 316–328.

47. Funiciello, R., Parotto, M., 1978. Il substrato sedimentario nell'area dei Colli Albani: considerazioni geodinamiche e paleogeografiche sul margine tirrenico dell'Appennino Centrale. Geologica Romana 17, 233–287.

48. Gambardella, B., Cardellini, C., Chiodini, G., Frondini, F., Marini, L., Ottonello, G., Vetushi Zuccolini, M., 2004. Fluxes of deep CO_2 in the volcanic areas of central-southern Italy.

Journal of Volcanology and Geothermal Research 136, 31–52.

49. Gaus, I., Le Guern, C., Pearce, J.M., Pauwels, H., Shepherd, T.J., Hatziyannis, G., Metaxas, A., 2004. Comparison of long-term geochemical interactions at two natural CO2-analogues:

50. Montmiral (Southeast Basin, France) and Messokampos (Florina Basin, Greece) case studies.In: The 7th Greenhouse Gas Control Technologies Conference (GHGT7), Vancouver, Canada, 5–9 September 2004.

51. Gerlach, T.M., Doukas, M.P., McGee, K.A., Kessler, R., 2001. Soil efflux and total emission rates of magmatic CO2 at the Horseshoe Lake tree kill, Mamoth Mountain, California, 1995–1999. Chemical Geology 177, 101–116.

52. Gianelli, G., Scandiffio, G., 1989. The Latera geothermal system (Italy): chemical composition of the geothermal fluid and hypotheses on its origin. Geothermics 18, 447–463.

53. Gibson, R.G., 1994. Fault zone seals in siliciclastic strata of the Columbia basin, offshore Trinidad. American Association of Petroleum Geologists Bulletin 78, 1373–1385.

54. Gray, M.B., Stamatakos, J.A., Ferrill, D.A., Evans, M.A., 2005. Fault-zone deformation in welded tuffs at Yucca Mountain, Nevada, USA. Journal of Structural Geology 27, 1873–1891.

55. Haynekamp, M.R., Goodwin, L.B., Mozley, P.S., Haneberg, W.C., 1999. Controls on fault zone architecture in poorly lithified sediments, Rio Grande rift, New Mexico: implications for fault-zone permeability and fluid flow. In: Haneberg, W.C., Mozley, P.S., Moore, J.C., Goodwin, L.B. (Eds.), Faults and Subsurface Fluid Flow. Geophysical Monograph 113. pp. 27–50.

56. Hinkle, M., 1994. Environmental conditions affecting concentrations of He, CO2, O2 and N2 in soil gases. Applied Geochemistry 9, 53–63.

57. Holloway, S., Pearce, J.M., Hards, V.L., Ohsumi, T., Gale, J., 2007. Natural emissions of CO2 from the geosphere and their

bearing on the geological storage of carbon dioxide. Energy 32, 1194–1201.

58. Hutchinson, G.L., Livingston, G.P., 1993. Use of chamber systems to measure trace gas fluxes. Agricultural ecosystem effects on trace gases and global climate change. ASA Special Publication, no. 55.

59. Hutchinson, G.L., Mosier, A.R., 1981. Improved soil cover method for field measurement of nitrous oxide fluxes. Soil Science Society of America Journal 45, 311–316.

60. IPCC, 2005. IPCC Special Report on Carbon Dioxide Capture and Storage. Prepared by Working Group III of the

61. Intergovernmental Panel on Climate Change. In: Metz, B., Davidson, O., de Coninck, H.C., Loos, M., Meyer, L.A. (Eds.), Cambridge University Press, Cambridge, United Kingdom, New York, NY, USA, p. 442.

62. IPCC, 2007. Climate change 2007: the physical science basis. Contribution of Working Group I to the Fourth Assessment Report of the Intergovernmental Panel on Climate Change. In: Solomon, S., et al. (Eds.), Cambridge University Press, Cambridge, United Kingdom/ New York, NY, USA, p. 996.

63. King, C.Y., King, B.S., Evans, W.C., Zang, W., 1996. Spatial radon anomalies on active faults in California. Applied Geochemistry 11, 497–510.

64. Le Nindre, Y.M., Pauwels, H., Rubert, Y., Shepherd, T.J., Pearce, J.M., 2006. CO2 trapping at geological time scale: geohistory of a natural analogue for long-term storage. In: The 8th

65. Greenhouse Gas Control Technologies Conference (GHGT8), Trondheim, Norway, 19–22 June 2006.

66. Lewicki, J.L., Birkholzer, J., Tsang, C.F., 2007a. Natural and industrial analogues for leakage of CO2 from storage reservoirs: identification of features, events, and processes and lessons learned. Environment Geology 52, 457–467, doi:10.1007/s00254-006-0479-7.

67. Lewicki, J.L., Brantley, S.L., 2000. CO2 degassing along the

San Andreas fault, Parkfield, California. Geophysical Research Letters 27 (1), 5–8.

68. Lewicki, J.L., Evans, W.C., Hilley, G.E., Sorey, M.L., Rogie, J., Brantley, S.L., 2003. Shallow soil CO2 flow along the San Andreas and Calaveras Faults, California. Journal of Geophysical Research 108 (B4), 2187, doi:10.1029/2002JB002141.

69. Lewicki, J.L., Hilley, G.E., Tosha, T., Aoyagi, R., Yamamoto, K., Benson, S.M., 2007b. Dynamic coupling of volcanic CO2 flow and wind at the Horseshoe Lake tree kill, Mammoth Mountain, California. Geophysical Research Letters 34 (L03401), doi:10.1029/2006GL028848.

70. Lewicki, J.L., Oldenburg, C.M., Dobeck, L., Spangler, L., 2007c. Surface CO2 leakage during two shallow subsurface CO2 releases. Geophysical Research Letters 34 (L24402), doi:10.1029/2007GL032047.

71. Lombardi, G., Mattias, P., 1987. The kaolin deposits of Italy. L'Industria Mineraria 6, 1–34.

72. Lombardi, S., Annunziatellis, A., Beaubien, S.E., Ciotoli, G., Coltella, M., 2008. Natural analogues and test sites for CO2 geological sequestration: experience at Latera, Italy. First Break 26, 39–43.

73. Lombardi, S., Annunziatellis, A., Ciotoli, G., Beaubien, S.E., 2006. Near surface gas geochemistry techniques to assess and monitor CO2 geological sequestration sites. In: Lombardi, S., Altunina, L.K., Beaubien, S.E. (Eds.), Advances in the Geological Storage of Carbon Dioxide. NATO Science Series, IV. Earth and Environmental Sciences, vol. 65. Springer, Dordrecht, The Netherlands, pp. 141–156.

74. Lombardi, S., Pinti, D.L., Rossi, U., Fiordalisi, A., 1993. 222Rn in soil gases at Latera geothermal field: a preliminary case history. Geologica Romana 29, 391–399.

75. Minissale, A., 2004. Origin, transport and discharge of CO2 in central Italy. Earth-Science Reviews 66, 89–141.

76. Morley, C.K., Wonganan, J.K., 2000. Normal fault

displacement characteristics, with particular reference to synthetic transfer zones: Mae Moh mine, northern Thailand. Basin Research 12, 307–327.

77. Morrow, C.A., Shi, L.Q., Byerlee, J.D., 1984. Permeability of fault gouge under confining pressure and shear stress. Journal of Geophysical Research 89, 3193–3200.

78. Nappi, G., 1969. Genesi ed evoluzione della Caldera di Latera. Bullettino del Servizio Geologico d'Italia 90, 61–68.

79. Nappi, G., Renzulli, A., Santi, P., 1991. Evidence of incremental growth in the Vulsinian calderas (central Italy). Journal of Volcanology and Geochemical Research 47, 13–31.

80. Nicoletti, M., Petrucciani, C., Piro, M., Trigila, R., 1979. Nuove datazioni vulsine per uno schema di evoluzione dell'attivita` vulcanica: il quadrante nord-occidentale. Periodico di Mineralogia 47/48, 153–165.

81. Oldenburg, C.M., Unger, A.J.A., 2004. Coupled vadose zone and atmospheric surface layer transport of carbon dioxide from geologic carbon sequestration sites. Vadose Zone Journal 3, 848–857.

82. Palladino, D.M., Agosta, E., 1997. Pumice fall deposits of the Western Vulsini Volcanoes (central Italy). Volcanology and Geothermal Research 78, 77–102.

83. Palladino, D.M., Simei, S., 2005. The Latera Volcanic Complex (Vulsini, central Italy): eruptive activity and caldera evolution. Acta Vulcanologica 17, 75–80.

84. Palladino, D.M., Valentine, G.A., 1995. Coarse-tail vertical and lateral grading in pyroclastic flow deposits of the Latera Volcanic Complex (Vulsini, Central Italy): origin and implications for flow dynamics. Journal of Volcanology and Geochemical Research 69, 343–364.

85. Pearce, J.M. (Ed.), 2004. Natural analogues for the geological storage of CO2. Final report of the Nascent project. British Geological Survey Technical Report, 122 pp.

86. Pearce, J.M., 2006. What can we learn from natural

analogues? In: Lombardi, S., Altunina, L.K., Beaubien, S.E. (Eds.), Advances in the Geological Storage of Carbon Dioxide. NATO Science Series, IV. Earth and Environmental Sciences, vol. 65. Springer, pp. 129–140.

87. Pettinelli, E., Beaubien, S.E., Lombardi, S., Annan, A.P., 2008. GPR, TDR, and geochemistry measurements above an active gas vent to study near-surface gas-migration pathways. Geophysics 73 (1), doi:10.1190/1.2815991.

88. Pfanz, H., Vodnik, D., Wittmann, C., Aschan, G., Batic, F., Turk, B., Macek, I., 2007. Photosynthetic performance (CO2- compensation point, carboxylation efficiency, and net photosynthesis) of timothy grass (Phleum pratense L.) is affected by elevated carbon dioxide in post-volcanic mofette areas. Environmental and Experimental Botany 61, 41–48.

89. Rawling, G.C., Goodwin, B.L., Wilson, J.L., 2001. Internal architecture, permeability structure, and hydrologic significance of contrasting fault-zone types. Geology 29 (1), 43–46.

90. Risk, D., Kellman, L., Beltrami, H., 2002. Soil CO2 production and surface flux at four climate observatories in eastern Canada. Global Biogeochemical Cycles 16, 69–80.

91. Sabatelli, F., Mannari, M., 1995. Latera development update. In: Proceedings World Geothermal Congress, Florence, 18–31 May 1995.

92. Sholtz, C.H., 1987. Wear and gouge formation in brittle faulting. Geology 15, 493–495.

93. Sibson, R.H., 1977. Fault rocks and fault mechanisms. Journal of the Geological Society of London 133, 191–213.

94. Sibson, R.H., 1996. Structural permeability of fluid-driven fault-fracture meshes. Journal of Structural Geology 18, 1031–1042.

95. Sibson, R.H., 2000. Fluid involvement in normal faulting. Journal of Geodynamics 29, 469–499.

96. Simei, S., Acocella, V., Palladino, D.M., Trigila, R., 2006. Evolution and structure of Vulsini calderas (Italy). Geophysics

Research Abstracts 8 (09302).

97. Sparks, R.S.J., 1975. Stratigraphy and geology of the ignimbrites of Vulsini Volcano, Italy. Geologische Rundschau 64, 497–523.

98. Stephens, J.C., Hering, J.G., 2002. Comparative characterization of volcanic ash soils exposed to decade-long elevated carbon dioxide concentrations at Mammoth Mountain, California. Chemical Geology 186, 301–313.

99. Stevens, S.H., Schoell, M., Ballentine, C., Hyman, D.M., 2004. Isotopic analysis of natural CO2 field: how long has nature stored CO2 underground? In: The 7th Greenhouse Gas Control Technologies Conference (GHGT7), Vancouver, Canada, 5–9 September 2004.

100. Tellam, 1995. Hydrochemistry of the saline ground-waters of the lower Mersey Basin Permo-Triassi sandstone aquifer, UK. Journal of Hydrogeology 165, 45–84.

101. Trigila, R., Walker, G.P.L., 1981. The Onano spatter flow, Italy: evidence for a new ignimbrite depositional mechanism. In: IAVCEI International Volcanology Congress, New Zealand.

102. Turbeville, B.N., 1992. 40Ar/39Ar ages and stratigraphy of the Latera caldera, Italy. Bulletin of Volcanology 55, 110–118.

103. Uehara, S., Shimamoto, T., 2004. Gas permeability evolution of cataclasite and fault gouge in triaxial compression and implication for changes in fault zone permeability structure through the earthquake cycle. Tectonophysics 378, 183–195.

104. Vezzoli, L., Conticelli, S., Innocenti, F., Landi, P., Manetti, P., Palladino, D.M., Trigila, R., 1987. Stratigraphy of the Latera Volcanic Complex: proposals for a new nomenclature. Periodico di Mineralogia 56, 89–110.

105. Wang, S., Jaffe, P.R., 2004. Dissolution of a mineral phase in potable aquifers due to CO2 releases from deep formations. Effect of dissolution kinetics. Energy Conversion and Management 45, 2833–2848.

106. Wells, A.W., Diehl, J.R., Bromhal, G., Strazisar, B.R., Wilson, T.H., White, C.M., 2007. The use of tracers to assess leakage

from the sequestration of CO2 in a depleted oil reservoir, New Mexico, USA. Applied Geochemistry 22, 996–1016.

107. Wibberley, C.A.J., Shimamoto, T., 2003. Internal structure and permeability of major strike–slip fault zones: the Median Tectonic Line in Mie Prefecture, southwest Japan. Journal of Structural Geology 25, 59–78.

108. Wilson, M., Monea, M. (Eds.), 2004. IEA GHG Weyburn CO2 Monitoring and Storage Project Summary Report 2000–2004. III. Petroleum Technology Research Centre, Regina, 273 pp.

109. Zhang, S., Cox, S.F., 2000. Enhancement of fluid permeability during shear deformation of a synthetic mud. Journal of Structural Geology 22, 1385–1393.

110. Zhang, S., Tullis, T.E., 1998. The effect of fault slip on permeability and permeability anisotropy in quartz gouge. Tectonophysics 295, 41–52.

111. Zhu, W., Wong, T.-F., 1997. The transition from brittle faulting to cataclastic flow: permeability evolution. Journal of Geophysical Research 102, 3027–3041.

112. Zhu, W., Wong, T.-F., 1999. Network modeling of the evolution of permeability and dilatancy in compact rock. Journal of Geophysical Research 104, 2963–2971.

113. Zoback, M.A., Byerlee, J.D., 1975. The effect of microcrack dilatancy on the permeability of Westerly granite. Journal of Geophysical Research 80, 752–755.

Chapter 6

Mitigation and Remediation Technologies and Practices in Case of Undesired Migration of CO$_2$ from a Geological Storage Unit— Current Status

J.-C. Manceau[a, d], D.G. Hatzignatiou[b, c, d], L. de Lary[a, d], N.B. Jensen[b, d], and A. Réveillère[a, d]

[a]BRGM, 3 av. C. Guillemin BP36009, F-45060 Orléans Cedex 2, France

[b]IRIS, N-4068 Stavanger, Norway

[c]University of Stavanger (UiS), N-4036 Stavanger, Norway

[d]CO2GeoNet, 3 av. C. Guillemin BP36009, F-45060 Orléans Cedex 2, France

ABSTRACT

One of the main objectives of operators and regulators involved in CO_2 geological storage activities is to ensure that the injected CO_2 will remain safely in the underground for a long period of time. Therefore, in addition to the screening and evaluation of the performance of a potential CO_2 storage site, risks of unwanted migration in the subsurface should be addressed and adequately managed. This can include the use of methods to mitigate those risks and ultimately to remediate potential adverse effects. This paper reviews the status of knowledge with regards to the mitigation and remediation technologies, from mature techniques adapted from other fields, such as oil and gas industry and environmental clean-up, to research topics offering potential new possibilities. Several categories can be defined: (1) interventions on operational or decommissioned wells to re-establish their integrity; (2) pressure/fluid management techniques for countering the leakage driving forces and/or removing the leaking fluids; (3) emerging technologies providing new mitigation opportunities for controlling undesired CO_2 migration; (4) techniques to remediate the impacts potentially induced by such a migration. This technical state of the art is completed by the actual practices in the emerging field of CO_2 geological storage established from the regulatory requirements and guidelines, and from the experience gained in existing storage projects over the world. This article concludes on important best practices stemming from this review and on future challenges in terms of research topics and operational needs.

INTRODUCTION

CO_2 permanent containment is one of the main concerns of operators and regulators implementing CO_2 geological storage. The existence of a low permeability caprock is viewed as a major element for a safe containment of CO_2 in the target storage formation (IEA-GHG, 2011a); as a result, any potential pathway is of major concern since it may allow buoyant CO_2 to migrate along and reach an overlying formation or be emitted at the surface, potentially impacting fresh water resources (as identified in IEA-GHG, 2011b) or sensitive stakes at the surface, respectively.

Undesired CO_2 migration out of the storage formation may occur through two different types of pathways:

- Engineered pathways, i.e., abandoned and operational wells (for injection, production or observation). Leakage can occur through these wells in case of (a) poor well completion (cementing) or plugging for the abandoned wells, (b) extreme over-pressurization near the injection well or at caprock, (c) chemical degradation of cement and corrosion of well completion equipment, and (d) well equipment failure;

- Natural pathways like faults or relatively more permeable zones in the caprock. These natural paths within the geological system may exist prior to the beginning of the storage operations or be created by the injection-induced processes (pressure-induced fracturing or reactivation, geochemical degradation of the caprock sealing efficiency); CO_2 migration out of the storage unit may also be caused by exceeding the caprock capillary entry pressure or translating the CO_2 plume along the formation bedding thus reaching the site's spill point.

Managing such risk scenarios is of vital importance for the full-scale deployment of CO_2 geological storage. Consequently a dedicated, site-specific mitigation/remediation strategy has to be designed and set up as a standard HSE practice. The storage safety is then guaranteed through an adequate site selection and characterization leading to the site choice where the evolution of the CO_2 plume and potential adverse impacts of the storage are judged acceptable. On this given site, a specific risk management process should be carried out to anticipate the potential deviations from this acceptable behaviour. The International Standard ISO 31000 (ISO, 2009) proposes a framework dedicated to the risk management: after having set the scope of the risk management (context establishment), a proper risk assessment shall ensure the selection of the site-relevant risk scenarios (risk identification), the estimation of the risk level through the computation of the consequences and of the likelihood of the scenarios (risk analysis) and the comparison with acceptable thresholds (risk evaluation). When the risk level is assessed as unacceptable, options are needed to lower the risk level; this last step is the risk treatment stage. Monitoring of the risk management process shall notably allow the detection and characterization of the potential deviations with respect to the expected evolution scenario and the efficiency assessment of the risk treatment measures. Risk communication during all various stages, with the stakeholders involved in the project, is essential for a transparent process.

This paper focuses on the risk treatment stage. This includes mitigation techniques used to avoid an impact to occur or reduce its magnitude. This could be achieved through a reduction of the likelihood of failure, through an action on the source of risks or through an intervention on the leakage pathway after the detection, evaluation and quantification of the leakage size, location and magnitude. In case an impact occurs, risk treatment also comprises remediation techniques applied to restore the environment.

To date this subject has not been fully addressed in the CO_2 geological storage literature despite the fact that it is currently receiving more attention from both industrial and scientific communities. This evolution may be linked with the new regulations on CO_2 geological storage that specify requirements on mitigation and remediation methods. The European Directive on the geological storage of carbon dioxide (EC, 2009) imposes, in the storage permit application, the inclusion of a description of measures to prevent significant irregularities as well as a proposed corrective measures plan. The United States Federal Requirements under the Underground Injection Control Program for Carbon Dioxide Geologic Sequestration Wells (US EPA, 2010a) require operators to submit an emergency and remedial response plan in the permit application. In Australia, the Offshore Petroleum and Greenhouse Gas Storage Act 2006 (Australian Government, 2011) and the associated regulations (Australian Government, 2012) require well operations management plan, which should include explanations on how the risks can be dealt with.

Given this growing interest, the objective of this paper is to supply an exhaustive review on techniques tested and implemented in other industries and contemplated to prevent, correct and remediate any undesirable CO_2 migration in the subsurface – second section – and on the actual practices currently followed in the emerging field of CO_2 storage, based on a survey of risk management conducted with the operators of several ongoing CO_2 and natural gas storage sites – third section. The state of the art produced in this study highlights several challenges in terms of existing and new technologies development and operational issues, which are developed as concluding remarks.

ESTABLISHED AND UNDER-DE-VELOPMENT MEASURES AND TECHNOLOGIES

Experience in the use of mitigation or remediation techniques is non-existent in the CO_2 geological storage field and the methods and techniques mentioned in the literature are mainly adapted from other domains/industries such as oil and gas industry or environmental clean-up. Some of these techniques have been tested and applied in several cases and under conditions that may be close to CO_2 geological ones, whereas some others are niche technologies but have been considered relevant to address challenges (depending on the CO_2 leakage type, location and rates, technical limitations of a given solution, environmental and cost issues) encountered in CO_2 sequestration applications. All of these technologies are gathered in this section, which should be considered as a state of knowledge of potential mitigation and remediation options, regardless to should they be standard and technically feasible at present time or innovative and under development.

The intervention on leakage through man-made pathways (well remediation) stems from the oil and gas industry experience and, in certain cases, mitigation operations can be considered as standard. However, in other cases the operator may not be able to rely on well engineering experience, and may have to turn to fluid management techniques or new breakthrough technologies for modifying leakage-path conductivity, fluid properties or CO_2 migration flow streams within the storage unit. This is true for most cases of leakage through natural pathways. In the case of impacting CO_2 migration, measures may be applied to remediate environmental impacts.

Accordingly, we divided our literature review according to four kinds of mitigation and remediation measures: interventions on wells, fluid management practices, breakthrough technologies and remediation measures on potential impacts.

Interventions on Leaking Wells

Existing production, injection, observation or abandoned wells are elevated risk elements to CO_2 storage projects since they can potentially act as leaking pathways (Nordbotten et al., 2005, Gasda et al., 2008 and Ide et al., 2006). Although current well closure and abandonment technologies appear to be sufficient to contain CO_2 at most sites (IEA-GHG, 2009), individual wells may suffer from a variety of factors that limit their integrity, including improper cementing or plugging, cement reactivity, casing/tubing corrosion, exposure to over-pressure and other failure conditions.

Leakage of a well can occur through the wellbore, the annulus between well tubing and casing, or on the outside of the casing. Gasda et al. (2004) list 6 different pathways for abandoned wells, namely (a) between casing and cement; (b) between cement plug and casing; (c) through the cement pore space as a result of cement degradation; (d) through casing as a result of corrosion; (e) through fractures in cement; and (f) between cement and rock. For active injection wells, packers also represent a possible leakage pathway.

CO_2 gas injected into a wet environment lowers pH and forms a corrosive environment for materials such as cement and steel. In addition, CO_2-enriched water may react with Portland cement used to isolate a well with the surrounding formations forming calcium carbonate. If wet CO_2 is still available the formed $CaCO_3$ can dissolve, thus degrading the cement. This degradation rate varies (IEA-GHG, 2009) and field evidence suggests that actual rates are low (Carey et al., 2007 and Crow et al., 2010). In addition, mechanical deformation of well cement and casing, due to poor cementing (Barclay et al., 2002) and or cement shrinkage (Ravi et al., 2002), may create important migration pathways at the interfaces between cement and casing or cement and caprock. Failure of cement sheaths can occur under stress caused by high injection pressures and/or temperature cycles and may result in the development of fractures. Finally, there is also a possibility of well deformation due to decompaction of the reservoir as the pressure increases during injection of CO_2. Though the reservoir rock and well or plug cement have similar mechanical properties, the casing steel has an elasticity modulus that is significantly higher and the strain level at the cement–steel interface may result in debonding and subsequently formation of micro-annuli at the interface. In the description of mitigation measures that follows, it is assumed that procedures for identification of CO_2 leak type and position in the well have been completed (IEA-GHG, 2007a).

With reference to the likely expense of a particular well operation, we refrain from including quantitative estimates due to large variations in the cost depending on the location, environment, country, well and formation characteristics, etc. However, based on in-house experience we give estimates as to the expected time needed to perform the operations and the factors controlling it.

Wellhead (and Welltree) Repair

The wellhead and welltree should be the first item to be checked prior to any in-depth leak detection investigation. Wellhead equipment, including valves, flanges, etc., can easily be inspected and repaired due to their above ground location (IEA-GHG, 2007a; for illustrations please refer for example toJørgensen, 2007). For subsea installations this may not be an equally simple investigation to perform and any repair will involve use of well service vessel and Remotely Operated Vehicle (ROV) (McGennis, 2001). Depending on the technical problem the leakage may be stopped and repair carried out at the sea bottom with the welltree in place. More severe damages may require that the welltree is removed and brought to the well service vessel or even onshore for repair. Removing the welltree requires that the well is securely plugged and perhaps even killed while the repairs are being carried out.

If wellhead has suffered severe damage and even deformation of its parts one may initially apply other methods such as video or 3D modelling photogrammetry inspection of the wellhead (Maccormick et al., 2011 and Sloan et al., 2011). Comparison of the 3D model with as-built drawings are subsequently used to design bespoke tools to ensure well integrity, customized cutting tools for removing damaged wellhead parts, and design of new replacement parts (e.g. gasket, flanges, etc.) making it possible to return the blowout preventer (BOP) and rest of the welltree to the wellhead.

Cost for operations depends on the damage and whether the mitigation requires a workover, i.e. if the well has to be killed by setting a plug to isolate the reservoir followed by filling the tubing string with a kill fluid. The cost for repair of the wellhead and/or welltree will have to be added.

Packer Replacement

Permanent packers have two locking systems with opposite directed cone-shaped surfaces; this way the packer can withstand larges forces in both upward and downward direction. Between the two locks there is a packer element made of elastic synthetic material that insures the sealing between casing and tubing (for illustrations please refer for example to Jørgensen, 2007). If annular pressure is lost while casing and tubing can be shown to be intact, it is likely a packer sealing element that is leaking. A leaking permanent packer is removed by use of a packer mill. Traditionally, removing the packer has been a two-step process where milling operation is followed by a retrieving operation pulling out remaining packer parts and recovering debris by flushing the well. More recently Haughton and Connell (2006) presented a method for simultaneously milling and retrieving permanent production packers that reduces considerably milling and trip time.

If the packer is of the retrievable type the packer and the tubing injection string can be retrieved from the well and replaced by a new packer. For mechanically set packers the packer and tubing are run into the hole together and it is attached/locked in place by rotation, the application of weight-loaded force, or a combination of both (IEA-GHG, 2007a and Jørgensen, 2007). Some packers can be pulled out and reset by ordinary completion units, however if a packer has begun to leak over time it is advisable to replace it rather than resetting it as a new leak may develop in future.

The removal and replacement of a production/injection packer will take about two weeks and involve a workover including killing of the well, welltree removal, tubing retrieval and packer removal.

Tubing Repair

If the leak is located in the injection tubing string, the tubing can be pulled out by a completion unit and the leaking joint be replaced. While the string is out of the well it is important to conduct a thickness and visual inspection to assess the state of individual joints, and possibly replace any part that shows signs of wear and tear. After inspection and repairs are completed the string can be run back into the well and pressure tested to ensure well integrity (IEA-GHG, 2007a and IEA-GHG, 2007b). Instead of pulling the tube out from the hole it may also be possible to apply an expandable casing patch (cf. Section 2.1.5). This can be done with the welltree in place and may therefore not require a workover. The cost will depend on the well length. If the tubing has to be pulled out and replaced, the cost will be comparable to packer replacement.

Squeeze Cementing

Squeeze cementing is applicable for repairing a faulty casing, within the cement or between cement and casing or rock. It may also be used to stop migration between separate zones of the reservoir. Squeeze cementing is normally performed at the time of running the casing however, it may be used as a mitigation method as well. Depending on the remediation need, squeeze cementing operations can be performed above or below the fracture gradient of the exposed formation, using high pressure squeeze and low pressure squeeze, respectively. Squeeze cementing operation may take up to five days including workover operations.

Low pressure squeeze technique is probably more efficient in placing a controlled amount of cement in a problem area of the well. The area is isolated by setting packers above and below. Pressure is achieved by pressuring up on the cement and allowing the cement to filter out on the formation creating a block in the annulus or in the fractures of the primary cement and casing. Once the cement slurry has hardened or dehydrated to a sufficient extent, no more fluid will be displaced. Excess cement in the well may have to be removed by milling. This method is the industry standard corrective for a loss of casing integrity.

If the casing is severely damaged the milling process may result in increased damage to the casing. Cirer et al. (2012) proposed to use an epoxy reinforced fibreglass (ERFV) pipe inside the well and pump the cement into the annulus between the casing and the ERFV. Since the ERFV does not contain cement it can be used to guide the milling tool and secure that the operation will not violate the restored well integrity. Squeeze cementing may take up to five days including workover operations depending on well condition, type and extent of repair.

In some cases, such as open-hole completions, scab casing could be more appropriate than squeeze cementing. In those cases a scab casing or liner with a smaller diameter than the wellbore is placed over the problematic interval. Cement is then forced into the annulus behind casing to seal off any leakage (Dillon and Billings,

1994). The advantage over simple squeeze cementing is a more reliable and longer-term solution to a leakage problem but at the cost of a reduced hole diameter.

Patching Casing

This method can be an alternative to squeeze cementing for repairing casing leaks. The patch may also be used to provide strengthening of corroded or otherwise weakened casing or completion equipment. The operation normally takes less than one week. In the case where it is more economical to replace the casing above and below a damaged section this can be done using expandable casing patch (tie-back) connections. This technique creates a hydraulic and gas-tight connection between the old and new casing; it also preserves the internal diameter of the well.

Many examples of successful application of the technique have been published (e.g. Chustz et al., 2005, Daigle et al., 2000 and Storaune and Winters, 2005). The use of expandable casing has over the past 10–15 years become a standard technique in the petroleum industry and is today capable of serving more than remediation of well integrity problems. Durst and Ruzic (2009) provide a comprehensive list of applications where expandable tubular facilitate improved well stimulation and the use of advanced well completion.

Swaging

If the casing has become deformed or collapsed into the well, due to external pressures, it can be restored to its previous use through swaging. This method involves the use of a (swaging) tool acting as a circular wedge forcing the tubing or casing walls out with its steel jaws as it is driven through the deformed or collapsed section of the well. The pressure exerted by the swage can exceed 50 tonnes/in^2. The swaging can also be used to expand a liner or tube and fit it to the inner wall of the well (see Section 2.1.5; IEA-GHG, 2007a).

Well Killing

Any leakage that cannot be mitigated through the installed welltree and BOP requires a full workover of the well. At all times a well is generally required to have at least two barriers between the reservoir and the surface. Hence before removing the BOP and welltree from the top of the well, the well has to be temporarily plugged. A plug is set to isolate the reservoir (the primary barrier) and the well is subsequently killed by filling it with a heavy kill-fluid (the secondary barrier) to avoid a blowout. Setting the plug will require a wireline (WL) operation or for horizontal or inclined wells a WL with a tractor or coiled tubing. Plugging operation may take 1–2 days for the WL operation. Afterwards removal of the plug and kill-fluid is an additional day's work.

Well Plugging and Abandonment (P&A)

The "aggressive" nature of CO_2 means that good practices must be followed in relation to well P&A. The requirements for P&A of wells are defined by national or local safety authorities who generally set standards for when and where to use cement plugs across underground sources of drinking water, across hydrocarbon production zones and open perforations in casing as well as squeezing cement into non-cemented, cased holes. In addition there are common requirements to have at least two barriers. The present requirements are to a large extent formulated to address plugging of oil and gas wells and variations between the different regulations may reflect regional geology or reservoir challenges (discussions of guidelines and regulations can be obtained in references such as IEA-GHG, 2007a, IEA-GHG, 2007b and Abshire et al., 2013; NORSOK D-010, Standard Norge, 2004; Guidelines for the Suspension and Abandonment of Wells, UK (Offshore Operators Association, 2012; Mining Regulations of the Netherlands WJZ02063603, Mijnbouvvet, 2003, etc.).

NORSOK D-010 has defined well barrier acceptance criteria for the function and type of well barriers. The function of a well barrier and plug can be combined if it fulfils more than one of the objectives, but at no time can a secondary well barrier be the primary well barrier for the same reservoir. A permanent well barrier should have the following properties: impermeable; long term integrity; non-shrinking; ductile – (non-brittle) – able to withstand mechanical loads/impact; resistant to different chemicals/substances (H_2S, CO_2 and hydrocarbons); and wetting, to ensure bonding to steel.

Based on the practice from the Norwegian Continental Shelf (NCS) a P&A workflow includes (a) planning, acquirement and mobilization of suitable vessel for the P&A operations, (b) kill the well and possibly remove the Christmas tree but not the BOP, (c) pull out the tubing and completion equipment; (d) perform a diagnostic logging to assess the conditions of the well; (e) plug the reservoir and potential cross-flow.(f) log, cut/pull intermediate casing and set extra plugs; (g) set top plug; (h) remove of upper part of surface casing, conductor and wellhead.

For CO_2 injection wells the P&A requirements need to address the very corrosive nature of CO_2 when it comes in contact with the cement and casing materials normally used for oil and gas wells. Randhol et al. (2007) propose new procedures and design for safe plugging of CO_2 injection wells. The possibility that casing corrosion will create channelling and that shrinkage or expansion of casing and cement can lead to formation of micro-annuli lead them to propose a list of five requirements that the sealing elements should provide and comply with: (a) multiple pressure barriers; (b) avoiding underground cross-flow between layers; (c) zero transmissivity; (d) chemically inert; and (e) provide sufficient bonding strength.

A chemical inert sealing material is not likely to exist so it is recommended that the requirement should be a material that has proven stability over time. Based on the studies of field cement samples by Carey et al. (2007) and Crow et al. (2010), oil well cements may be suitable and fitting as a long lasting sealing material.

The requirement of sufficient bonding strength may also be difficult to deal with in practical terms as there exists no useful definition of bonding strength. It is therefore not possible to define a value that can be said to be "sufficient". If micro-annuli form, they are likely to form as a result of an excessive parting force.

Managing Abandoned Wells

According to Friedmann (2007) and Nicot (2009) abandoned wells are more likely to leak due to (a) improper abandonment practices, (b) abandonment procedures that have not been followed properly, or (c) well-abandonment not designed for long-term protection (e.g. well seal failure). Though mapping and assessing of potential leakages from abandoned wells should be addressed during the storage evaluation process prior to injection, some wells may not be discovered until after injection has started or even finished. Depending on the condition of the particular well, the measures needed for mitigation and remediation must be selected accordingly to ensure adequate seal against the continued fluid and CO_2 migration. Operation on the well may require setting up a rig above the location of the leaking well and applying any necessary method to stop the leak. Recently, a global service company reported (2012[1]) that they successfully performed rigless P&A operations in the North Sea on several wells. Finally, the well must be replugged according to national legislation. Apart from the typical onshore wells, a special mentioning should be made to subsea offshore wells or offshore wells which have been cut off below the seabed and may require a relief well for addressing potential leaks due to temporarily abandonment, ageing materials, or faulty permanent abandonment. Locating old, abandoned wells underneath the seabed in itself may pose a challenge as may removal of the cement cap that would have been placed over the cut casing top. New methods and tools focussing on reduced costs for downhole P&A operations are available (e.g. Abshire et al., 2012).

Stopping Surface Blowout

If the main barrier (e.g. tubing, packers, or the primary seal) is lost this in itself does not result in a blowout. It should be possible to avoid breaching of any secondary barriers by reducing injection and/or remediating the migration within the wellbore. In many respects a CO_2 blowout can be considered similar to hydrocarbon blowout. Skinner (2003) reports on three events of CO_2 blowouts and concludes that preventive measures such as regular inspection and maintenance of the Blow out Preventer Equipment (BOPE), and installation of additional BOPE on suspect wells are the key elements in reducing probability of blowouts from CO_2 injection wells.

Stopping a blowout requires killing the well through the injection of heavy fluids. The weight of kill fluid will form a new seal hindering CO_2 from flowing out from the well. This measure is relatively quick to apply and may at best be carried out within hours or days.

If the reason for the blowout is caused by missing or damaged wellhead or tree the situation is complicated by the cold and toxic conditions. The specific circumstances and risks should be assessed before any decision is taken to attempt stopping the blowout (ScottishPower CCS Consortium, 2011). If special tools need to be designed and fabricated or transported to the site, this results in longer response time before the leakage is under control, possibly day to weeks.

If one cannot access the wellhead it can be a solution to bring in a rig and drill a relief well to lower the pressure and bring the leakage through the damaged well under control. The design of the relief well is similar to that for a hydrocarbon well. The intersection point is likely to be relative deep, i.e. near the lowest casing within the reservoir section. It is important that the formation does not fracture under the pressure of the escaping fluids (ScottishPower CCS Consortium, 2011).

The cost of drilling a new relief well is dependent on length and/or time needed to stop the leakage and take control over the surface blowout on the first well. It is also dependant on the offshore/onshore location of the well. Table 1 summarizes each well intervention technique with precisions on the aim of these measures.

Table 1: Summary of proposed techniques for mitigation of leakages from wells

Main Objective	Mitigation techniques
Wellhead and welltree inspections and repairs	For above ground installations inspections/repairs of valves, flanges etc. are relatively easy as they are accessible (IEA-GHG, 2007a). Subsea cases require use of remotely operated vehicles (ROV). Depending on the severity and nature of the problem the well may be securely plugged while operations are performed (McGennis, 2001, Maccormick et al., 2011 and Sloan et al., 2011).
Replacement of a leaking packer	Retrievable packers can be removed in a wireline operation (IEA-GHG, 2007a and Jørgensen, 2007). Permanent packers are removed by milling (Haughton and Connell, 2006).
Replacing or repairing leaking or damaged tubing	Pull and replace string (IEA-GHG, 2007a). Repair using an expandable patch (Chustz et al., 2005, Daigle et al., 2000, Durst and Ruzic, 2009 and Storaune and Winters, 2005).
Repairing a leaking casing or preventing migration between separate reservoir zones through annulus	Squeeze cementing (IEA-GHG, 2007a and Cirer et al., 2012). Scab casing (Dillon and Billings, 1994).
Replacing or repairing leaking or damaged casing	Repair casing using an expandable patch (Chustz et al., 2005, Daigle et al., 2000, Durst and Ruzic, 2009 and Storaune and Winters, 2005).
Repairing deformed or collapsed casing	Swaging (IEA-GHG, 2007a).
Shutting down an injection well or repairing leaking old abandoned well	The requirements for plugging and abandonment (P&A) of wells are defined by national or local safety authorities – cf. IEA-GHG (2007a) for examples from North America; NORSOK D-010 (2004) for the Norwegian Continental shelf, Guidelines for the Suspension and Abandonment of Wells (2009) for the UK sector, Mining Regulations of the Netherlands WJZ02063603 (2003). Design and procedures for plugging and abandoning (P&A) of CO_2 injection wells are described by Randhol et al. (2007).
Stopping a surface blowout	Well killing and drilling of a relief well (Skinner, 2003; ScottishPower CCS Consortium, 2011).

Fluid Management Practices

As described in the previous section, there is a portfolio of repair or plugging options available for leaking wells, which are a priori accessible. For some migrations cases such as caprock sealing defects including faults, fractures and high permeability areas, few measures can directly target the leakage pathways. In these cases, the mitigation measures therefore aim at countering the forces responsible for the CO_2 migration and several options, all implying fluid managements, have been proposed in order to prevent or minimize CO_2 migration.

Fluid management solutions would be employed to (a) temporarily or permanently arrest the pressure increase or decrease the pressure in the storage aquifer, locally or globally, (b) create a pressure barrier in the overlying geological strata to prevent or minimize CO_2 leakage, (c) back-produce injected CO_2 either locally or globally, and (d) enhance non-structural trapping mechanisms.

Pressure Relief in the Storage Formation

Natural processes of brine and rock compression, as well as dissolution of CO_2 into formation brine through density-driven convection, will naturally decrease the pressure build-up in the formation, as presented by the IEA-GHG (2010a). The same study, however, notes that weak density difference between CO_2-saturated and unsaturated brine could hinder this process. Operational choices are therefore necessary for reducing the level of reservoir pressurization in case of an abnormal behaviour. Stopping the CO_2 injection may be considered as a mitigation technique if the pressure relief in the storage formation is sufficient for reducing leakage, or preventing the CO_2 plume from reaching a leakage pathway. Since this option might not be sufficient for preventing CO_2 leakage outside the storage reservoir, accelerated and enhanced strategies, such as drilling new injection wells, producing at the injection well or extracting brine at a distant location, may also be considered for application. Le Guénan and Rohmer (2011) for instance investigated and compared means of controlling the overpressure for a CO_2 storage scenario applied to the Paris basin.

It should still be noted that in the case where the over-pressurization has created a new leakage pathway through, e.g., fault reactivation and hydraulic fracturing, these created cracks and reactivated faults may not totally close with the sole pressure relief. This mitigation option can therefore not be sufficient for stopping a CO_2 leakage if the mobile plume reaches these areas, due to buoyancy effects and viscous forces. This last case is an argument for considering pressure control strategies in the injection plan, and not solely as a mitigation option. Estimation of pressurization and brine displacement over time in the case of CO_2 storage, as well as potential effects on caprock and fault integrity have been reviewed in IEA-GHG (2010b). Different injection strategies are presented in IEA-GHG (2010a) notably with the purpose of limiting the overpressure created by the CO_2 injection, and therefore to avoid creating leakage pathways through e.g. fault reactivation. Lindeberg et al. (2009), Hatzignatiou et al. (2011), Bergmo et al. (2011) andBirkholzer et al. (2012) present and evaluate as well injection strategies, using brine production wells, increasing the number of injection wells and/or using horizontal wells. Additionally, a recent study dedicated to formation water extraction issues (IEA-GHG, 2012), assesses the potential benefits in terms of pressure management and discusses different brine disposal options on different case studies (Ketzin – Germany, Zama – Canada, Gorgon – Australia, Teapot Dome – USA).

Hydraulic Barrier

Hydraulic barriers are used as a preventive or corrective measure in pollution engineering. This technology consists in injecting (or producing) water to locally modify the hydrogeology and protect the drinking water against saline brine intrusion, which is one of the most widespread forms of fresh groundwater pollution in coastal areas (Parrek et al., 2006; US EPA, 1999). Several authors have considered adapting this method to the case of a CO_2 leakage from the storage reservoir to an overlying aquifer that is not deemed a sensitive asset. Injecting brine into the overlying aquifer will increase the pressure just above the leak to counter-balance the CO_2 buoyancy and the storage reservoir over-pressurization that are driving this leakage (Benson and Hepple, 2005 and IEA-GHG, 2011b). Implementing a hydraulic barrier requires the consideration of many operational and strategic issues: delays, costs and technical feasibility of re-using a former injection well or drilling a new one, levels of induced over-pressurization to avoid reactivation of existing faults and fractures widening or even creation of new ones, availability of brine, efficiency of the injection, and measure rate of response. Most of these issues might hinder the applicability of hydraulic barrier and restrain its design. Réveillère and Rohmer (2011) evaluate this applicability, both as corrective and preventive measures, and conclude that the distance from the injection well to the leakage appears as the most critical parameter. Hydraulic barrier may be efficient if applied in the immediate vicinity of the leakage plume; however, it may be an impractical solution at long distances since it requires long injection periods to be efficient.

CO_2 Plume Dissolution and Residual Trapping

In Situ Enhancement of Dissolution and Residual Trapping

Enhancing dissolution and residual trapping may be considered as a remediation option both for the injected CO_2 plume in the storage reservoir and/or a secondary accumulation in an overlying aquifer. The method relies on a brine flow over the CO_2 plume, which will enhance these two CO_2 trapping modes. The brine flow may be natural, i.e., due to groundwater flow - passive remediation - (Juanes et al., 2010), but active remediation options have been also discussed for trapping small leakage plumes (Esposito and Benson, 2012) or the injected CO_2 plume (Leonenko and Keith,

2008, Nghiem et al., 2009, Qi et al., 2009 and Manceau et al., 2011). However, the latter case requires large brine injection flow rate and induces overpressure, which raises several issues including the geo-mechanical integrity of the reservoir and potential brine leakages. This measure has been studied both as a design or mitigation option. Approximate analytical solutions (Juanes et al., 2010 and Manceau and Rohmer, 2011) or numerical models with implemented residual trapping modules (e.g. Doughty, 2007) have been proposed to estimate the efficiency of the natural or active trapping of the CO_2 plume at large scale.

Ex Situ CO_2 Dissolution and Saturated Brine Injection

Alternatively to the techniques aiming at enhancing non-structural trapping in situ, an option is to use surface or ex situ dissolution in order to store dense CO_2-saturated brine. Some authors have indeed proposed the use of this extraction/CO_2-dissolution/injection process whereby the saline aquifer brine is extracted via production wells, the captured CO_2 is dissolved into the extracted brine on the surface using high pressure/temperature mixing vessels and the CO_2-laden brine is re-injected into the storing formation (Leonenko and Keith, 2008, Burton and Bryant, 2009, Eke et al., 2011a, Eke et al., 2011b and Tao and Bryant, 2012). This method is also assessed in IEA-GHG (2012) for several case studies. The technique has been proposed as a storage design by these authors, but not as a corrective measure; it may therefore be considered as a preventive measure, or as major change of the injection strategy.

Although the extraction/CO_2-dissolution/injection process appears to be an attractive solution, it may be hampered due to the several reasons: (a) reservoir heterogeneities affecting the injected CO_2-laden brine movement into formation; (b) increased pressure regimes near the injectors; (c) decreased pressure regimes near the extractors; (d) the required large number of injection/production wells to enable the process and associated costs; (e) added costs required by the surface facilities; (f) need to optimize wells location; (g) possible near-well mineralization further reducing well injectivity or mineral dissolution that may threaten the reservoir integrity (André et al., 2007); and (h) deployment difficulties both onshore and offshore.

CO₂ Back Production

CO_2 storage is planned to be permanent. The European Commission Directive on CO_2 geological storage (EC, 2009) for instance states in the first Article, second paragraph, that *the purpose of environmentally safe geological storage of CO2is permanent containment of CO2*. Nevertheless, the back production of stored CO_2 might be considered as a remediation measure, if the site is less suitable than anticipated (Benson and Hepple, 2005).

Theoretically, all stored CO_2 in the formation can be back produced, except the CO_2 that is stored in the form of mineral trapping. However, as pointed out by Rohmer et al. (2009), the achievable back production ratio in real sites is limited by the complex and heterogeneous nature of the geological storage, which is in addition partially known and where various phenomena occur with sometimes very large time scales. Moreover, partial or total back production of the injected CO_2 has not been tested yet in CO_2 geological storage sites, and only few studies directly address this question. However, this is technically very similar to a leak in an open well, which has been the focus of many studies in the CO_2 geological storage literature using both analytical and numerical models (Pawar et al., 2009, Nordbotten et al., 2009 and Humez et al., 2011). It is also similar to the production of CO_2 from natural CO_2 reservoirs, or to natural gas (possibly from CO_2-rich reservoir) production. Gaseous CO_2 back production will therefore face similar challenges to natural gas production, and will benefit from the experience of this industry (number and positioning of wells, benefits of horizontal wells, coproduction of brine, etc.). The sole study directly considering large-scale back production of stored CO_2 is by Akervoll et al. (2009) who assessed the back production of CO_2 after 15 years of storage in the Utsira formation (Norway);

they showed that 47.7% of the injected CO_2 can be produced within 7 years of production through a single horizontal well. Back production of small leakage plumes has also been studied by Esposito and Benson (2012); the authors showed a better removal of small vertical plume relatively to large thin plumes of CO_2 (gravity tongues) at the top of the aquifer, which also induces the co-production of large quantity of brine that has to be managed.

As a summary of this section, each fluid management technique is reported in Table 2 with precisions on the aim of these measures.

Table 2: Summary of proposed fluid management techniques for mitigation of CO_2 migration in the subsurface

Main objective	Mitigation techniques
Decrease the pressure in the storage formation	Pressure control strategies (preventive measure): Lindeberg et al., 2009, Hatzignatiou et al., 2011,Bergmo et al., 2011, IEA-GHG, 2010a, IEA-GHG, 2010b and IEA-GHG, 2012. Stopping the injection, production at the injection well, fluids extraction from additional wells (corrective measure): Le Guénan and Rohmer, 2011.
Counteract the buoyancy and pressure gradient driving the CO_2 migration	Hydraulic barrier: Benson and Hepple, 2005, IEA-GHG, 2011a, IEA-GHG, 2011b, Réveillère and Rohmer, 2011 and Réveillère et al., 2012.
Avoid migration of buoyant CO_2 by enhancing dissolution and residual trapping	In situ enhancement of trapping modes due to natural processes (groundwater flow) or to brine injection, in the storage reservoir or in overlying aquifers (in case of secondary accumulation):Leonenko and Keith, 2008, Nghiem et al., 2009, Qi et al., 2009, Manceau et al., 2011, Juanes et al., 2010, Manceau and Rohmer, 2011 and Esposito and Benson, 2012. Ex situ CO_2 dissolution and saturated brine injection: Leonenko and Keith, 2008, Burton and Bryant, 2009 and Eke et al., 2011a and 2011b; Tao and Bryant, 2012.
Partial removal of injected CO_2	Partial CO_2 back production from the injection aquifer or from overlying aquifers (in case of secondary accumulation): Benson and Hepple, 2005, Rohmer et al., 2009 and Akervoll et al., 2009.

Breakthrough Technologies

Specific requirements for CO_2 geological storage (e.g., time duration, fluids mobility and reactivity, storage site pressurization) coupled with the continuous improvement of existing well intervention solutions and current advances in emerging technologies provide new opportunities for controlling unwanted CO_2 leakage out of the storage unit. Existing and breakthrough technologies for mitigating an undesired migration of the CO_2 plume include the use of (a) conventional Portland and geopolymer cement to isolate well/formation communication behind casing, (b) foams and gels to reduce CO_2 mobility and isolate conductive flow paths, (c) nanoparticles and biofilms to enhance the sequestration of CO_2 and reduce/eliminate any potential risk for CO_2 leakage.

The potential application of some of these techniques depends strongly on the location of the undesired CO_2 migration and the leakage severity. Depending on the storage formation and leakage-path properties, some of these techniques may serve as short- to intermediate-term solutions until a more permanent one (e.g., a sidetracked or new relief well is drilled in case of a major/catastrophic leakage which will serve to permanently isolate the source of leakage) can be placed to address long-term solutions.

Conventional Portland and Geopolymer Cement

The use of CPC (Conventional Portland Cement) as an isolation agent to control behind-casing/formation fluid communication with the reservoir formation and provide zonal isolation and integrity of wells has been an issue for a long time in the oil and gas industry (see for example Deremble et al., 2010, Loizzo and Duguid, 2006 and Gasda et al., 2004). The cement's ability, in any form, to penetrate into the formation matrix for distances larger than few inches is very limited for applied pressures below formation parting pressure. This fact, in addition to cement shortcomings during setting, makes the CPC a non-attractive solution for addressing larger scale (even close to wellbore) formation heterogeneities or well completion deficiencies. Over the years, several researchers have attempted to rectify this weakness of the CPC system to provide full isolation of the formation/behind-casing annulus and identify remedies to address fluid communication between subsurface strata and wellbore (see for example Loizzo et al., 2011, Watson and Bachu, 2009, Barlet-Gouedard et al., 2006, Talabani et al., 1993a and Talabani et al., 1993b). In addition, compared to traditional oil, gas or water wells, CO_2 storage context requires the ability of the materials used for the well completion to maintain a chemical resistance to supercritical CO_2 over a significant long time period.

Geopolymer displays a relatively higher strength, excellent volume stability, as well as better durability and resistance to acids when compared to CPC. Geopolymer has also a strong resistance to corrosion, lower manufacturing cost, 50% lower energy requirements, and 90% lower CO_2 emission during manufacturing; therefore, it could be a viable option to replace CPC in CO_2 sequestration well cementing (Davidovitch, 2005 and Nasvi et al., 2012).

Foams and Gels

Surfactant-stabilized foams are generally used for mobility control in gas-based enhanced oil recovery processes (see for example, Schramm, 1994, Rossen, 1996 and Hatzignatiou et al., 2014a). The use of foams as conformance control agents has found a limited use in the oil industry up to now. Their complex nature, the difficulty in controlling their strength in situ in the formation, and the limited use in economically viable conformance control applications have hindered them from becoming the first-choice solution. However, the use of new technologies, such as nanotechnology, could lead to an improved stability foam for conformance control application (Sydansk and Romero-Zerón, 2011 and Yu et al., 2012).

Foams, polymer or inorganic gels have been traditionally used in the oil industry to counteract production of unwanted fluids (water and/or gas) and also divert injected fluids into formation regions which have been poorly swept, thus containing significant amounts of mobile oil. The selection, design and deployment of the appropriate mobility-controlled agent are type specific and require, among others, the proper characterization of the storage site (including geological structure and geometry, formation characteristics, formation water salinity, formation pressure and temperature, pH level, etc.) as well as the CO_2 leakage location, type and size.

Deployed gels aim to reduce the permeability of an existing fluid conduit, thus reducing/controlling the leakage of a high mobility fluid such as that of the stored CO_2. A high-conductivity pathway could be located near an injection well or away from it. The injected chemicals, in general, have good injectivity enabling them to be deployed at distances away from an injection well. Once in place, an appropriate technique is employed, dependent on the deployed system that triggers the gelation process which leads to the development of a gel-like system that reduces significantly the permeability of the leakage pathway. The placement of the original fluids is also technology dependent and is also affected by the type and completion of the existing well. "Surgical" placement of these fluids, yielding the best possibilities for controlling unwanted fluids migration, may require the use of a sidetracked or dedicated slim-hole well. This itself not only increases the cost for controlling a potential CO_2 leakage, but it is also time consuming which means that CO_2 continues to leak outside of the storage unit with undesirable consequences at either one or both the subsurface and surface environments.

Polymer- polyacrylamides (PAM) or biopolymers-macromolecules are linked together by crosslinkers (normally metal ions, metallic complexes or organic agents) forming a viscous gel in the formation. Some issues related with polymers are gelation control, adsorption and ability for deep penetration, because of the inherent high viscosities. Several types of polymer-based gels such as movable gels, pH-sensitive polymers, BrightWater, microball, preformed particle gel, etc., have appeared in the literature in the last few years (for a summary see for example Sheng, 2011 and

Sydansk et al., 2005). Inorganic gels such as silicate gels have also resurfaced and been evaluated for applications to control unwanted fluids in oil-produced reservoirs (see for example Stavland et al., 2011a, Stavland et al., 2011b, Skrettingland et al., 2012 and Hatzignatiou et al., 2014b). In particular, the use of inorganic silicate solutions appears to be promising since CO_2 could be used as a gelation agent and to accelerate or control the injected system gelation.

The effectiveness of silicate gels in controlling the flow of unwanted fluids is due to their (a) deep penetration that can be achieved into the treated zone because of the low initial (i.e., prior to gelling) silicate fluid viscosity, (b) good thermal and chemical stability, (c) low cost, (d) environmental friendliness, and (e) easy "removal" in case of an unexpected deployment failure (Lakatos et al., 1999). Several researchers have experimented with silicate gels to enhance the formed gels' chemical and thermal stability, durability and strength by utilizing polymers or other chemicals (see for example, Burns et al., 2008 and Lakatos et al., 2009).

In general, the use of gels for controlling the movement of CO_2 within the structure or leakage out of the storage unit should be viewed as time-dependent solution. Issues such as gel strength, durability, dehydration, temperature, acid and bacterial resistance, etc. are of concern and the appropriate system should be sought after, designed and implemented to enhance the properties and duration of the deployed gel system for isolating a developed CO_2 leakage. The impact of the acidic environment in a CO_2 sequestration process needs to be evaluated and the impact of pH reduction on the gels' gelation time, strength and stability needs to be quantified in order to establish safe application boundaries.

Nanoparticles

Nanoparticle-based applications for mitigating undesirable CO_2 leakage from the storage formation have recently appeared in the literature.

Nanoparticle Use in Foams

Yu et al. (2012) proposed the use of a nanoparticle-stabilized supercritical CO_2 foam, which was successfully generated in a tube. The authors also observed that an increase of the brine salinity and temperature leads to a reduction of the CO_2 foam, whereas when pressure increases more CO_2 foam was generated. Yu et al. (2012) concluded that the addition of surfactant to the nanosilica dispersion improved CO_2 foam generation. This finding is encouraging for developing strong and easy-to-develop foams for practical applications for oil and gas as well as carbon sequestration applications.

Nanoparticles for Mobility Control

DiCarlo et al. (2011) reported measurements of flow pattern and in situ saturations observed when n-octane displaced brine that contained dispersed surface treated silica nanoparticles. The authors reported more uniform displacement fronts and delayed breakthroughs compared to a control displacement with no in situ nanoparticles. This finding along with pressure measurements which were consistent with the generation of a viscous phase (emulsion) suggests that a nanoparticle stabilized emulsion formed during displacement suppressed the viscous instability.

Nanoparticle Use in Silicate Gels

Lakatos et al. (2012) investigated the use of nanoparticles to improve the stability and flexibility of resulting silicate gels. The authors conducted laboratory studies to investigate the gelation mechanism, rheological properties, flow behaviour and nanoparticle type and size on the resulting silicate gels. They concluded that SiO_2 nanoparticles are compatible with silicate solutions and the stability of the resulting SiO_2/silicate system depends significantly on the size and concentration of nanoparticles: the smaller the size, the more stable and the higher the concentration and the less stable the resulting system. In addition, as the concentration of nanoparticles increases, the viscosity of the SiO_2/silicate system increases too whereas, for a given catalyst concentration and temperature conditions, the setting time is reduced. These results clearly indicate that the use of nanoparticles could be used to yield even better and more efficient silicate systems for field applications to control, limit or even eliminate potential CO_2 leakage through heterogeneities in the caprock and/or formation.

Nanoparticle Use in CO_2 Sequestration

Addition of nanoparticles to injected CO_2 has also been proposed by Javadpour and Nicot (2011) to enhance CO_2 storage and reduce CO_2 leakage risks in deep saline aquifers by increasing the density contrast between the CO_2-rich brine and the resident brine. According to the authors, the addition of nanoparticles will decrease the instability onset time and increase convective mixing due to CO_2 diffusion into the resident brine. Both metallic nanoparticles and depleted uranium oxide were proposed to reduce CO_2 leakage risks. The authors assumed that the nanoparticles do not adsorb onto the rock surface or impair the formation permeability. The use of waste materials, depleted uranium oxide nanoparticles, and their inherent leakage risk is an issue that needs to be addressed. Singh et al. (2012) conducted a simulation study to investigate the use of nanoparticles to enhance CO_2 storage in deep saline aquifers by expediting convective mixing and decreasing the CO_2 buoyancy flow. Based on their numerical results, the authors concluded that the injected nanoparticles-CO_2 plume dissolves deeper and moves less laterally than the normal (i.e., without nanoparticles) CO_2 plume. Therefore, the faster mixing and decrease CO_2 buoyancy could reduce the chances for CO_2 leakage through the caprock and also obviate some of post-CO_2 injection monitoring costs.

Biofilms

Biofilms have been proposed as means to control the spread of, and treat, a contaminant plume in subsurface formations (Cunningham et al., 2003) and for helping to prevent a leakage of stored supercritical CO_2 through the caprock by blocking leakage pathways (Cunningham et al., 2009 and Mitchell et al., 2009). According to the authors, this could be achieved by injecting biofilm-forming organisms and growth nutrients and controlling the spatial extent and mass of the biofilm. The created biofilm could also protect well cement from an "attack" by the CO_2-rich brine. Coreflood experiments demonstrated the reduction of permeability with the biofilm generation in the porous medium, and Field Emission Scanning Electron Microscope (FESEM) images displayed clearly the mineral surfaces with the assemblage attached to the mineral surfaces at the end of the experiment (Mitchell et al., 2009).

Mitchell et al. (2010) and Cunningham et al. (2011) proposed an "engineered biomineralization" which could potentially create a long-term stable low permeability zone. Biofilms and biominerals were envisioned to (a) enhance CO_2 structural trapping by pore clogging and CO_2 leakage reduction, (b) result in biofilm-enhanced mineralization of carbonate minerals (i.e., mineral trapping), (c) produce biofilm-enhanced solubilization of CO_2 (solubility trapping), and (d) improve protection of injection well casing.

The main issue of this technology is the inability to effectively control the biofilm generation and growth as well as its proper placement within the formation to the locations of the highest risk of a CO_2 leakage or of an already developed leakage. In general, the presence of H_2S-generating sulphate-reducing bacteria may create localized corrosion of the well casing and adverse interaction with its cement used for zonal isolation. Therefore, appropriate means to protect the well (casing and cement) against the potential presence of H_2S should be a concern that needs to be addressed when biofilms are further developed and fine-tuned to CO_2 storage settings.

Remediation Measures on Potential Impacts

This section is devoted to remediation of the impacts of CO_2

migration, should these impacts occur. A storage site would be selected to minimize risks to the environment; if a CO_2 leakage from the reservoir would occur, it could be from a discrete point source (e.g. leakage from abandoned well) leading to localized impacts (IEA-GHG, 2007b). However, in case of an uncontrolled leakage, impacts on the environment could be larger (West et al., 2005). The potential compartments, which may be impacted by an undesired CO_2 migration, are considered to be: groundwater aquifers (confined or unconfined), the unsaturated zone, and surface assets (including human health, ecosystems and other activities) (IEA-GHG, 2007a and IEA-GHG, 2007b).

Remediation measures specifically dedicated to CO_2 storage impacts are poorly documented. One of the most important gaps to be filled to enable the design of appropriate measures seems to be the precise knowledge of the leakage mechanisms and associated impacts. To our knowledge, no remediation action following CO_2 leakage after geological storage has ever been implemented mainly due to the absence of established impacts. Furthermore, no field scale experiment is currently available to assess whether or not a measure could be relevant for the CCS domain. Consequently, literature is mainly based on modelling or analogies with other pollutants and thus, the degree of maturity of remediation measures in the CCS field is low. Moreover, for the presented measures, costs and time for complete removal are generally highly site-dependent and no information is specifically available in this area in the CCS field.

Many measures suggested in this section rely on the possible analogy between CO_2 and VOCs (Volatile Organic Compounds), for which extensive remediation literature is available, as discussed notably byRohmer et al. (2010) and Zhang et al. (2004). At surface conditions, CO_2 presents similarities with VOCs due to its density (1.81 kg/m^3 at 25 °C, the same order of magnitude with VOCs), viscosity (intermediate value between water and air) and volatility (vapour pressure of gaseous CO_2 reaching the value of 58.5 × 10^5 Pa at 20 °C). However, despite these similarities, two main differences between VOCs and CO_2 should be considered: (a) CO_2

is harmless at low concentrations whereas numerous VOCs could be deleterious to human health and the environment at low dose, (b) in case of an unexpected leakage from a geological storage the CO_2 may migrate upward (from reservoir to surface) whereas during usual pollution by VOCs, due to leaks or spills, the pollutant migrates mainly downward or laterally.

Remediation Techniques for Impacted Groundwater

The main impacts to groundwater to be remediated are: accumulation of gaseous or dissolved CO_2; acidification of aquifers; contamination from the injected CO_2 stream, or from displaced or released species due to the CO_2–fluid–rock interactions (IEA-GHG, 2011b).

Monitored Natural Attenuation (MNA)

This remediation measure is commonly used for pollution by organic pollutant (petroleum related products, chlorinated solvents, etc.) and under certain conditions for inorganic contaminants (US EPA, 1999). It relies on in situ natural physical, biological, chemical processes to reduce the mass, toxicity, mobility, volume, or concentration of contaminants in soil or groundwater (US EPA, 1999 and Khan and Husain, 2003). An adequate monitoring plan has to be set up to check its effectiveness and to ensure that the risks due to pollution are appropriately managed. Cost and time to perform preliminary studies could be higher than for active remediation techniques, however, long term costs may be lower. The time needed for remediation is site-specific and generally considered longer than for active remedial measures. A reasonable time frame for natural attenuation is few years and should not be higher than 30 years (US EPA, 2004 and Carey et al., 2000). The efficiency of natural attenuation is extremely dependent on site specific conditions (Colombano et al., 2010). Natural dissolution, dilution and mineralization of CO_2 (or other associated substances) through diffusion or advection processes (due to the groundwater flow for instance) could be a way to reduce the contaminant concentrations at an acceptable pace (Benson and Hepple, 2005). Based on modelling of intrusion of CO_2 on a glauconitic-sandstone aquifer, Vong et al. (2011) showed that concentrations of dissolved Pb and Zn and Cd decrease due to natural attenuation when leakage is stopped. Natural transformation of contaminants to another less toxic product, reduction of mobility or bioavailability might be other beneficial effects of natural remediation in case of pollution by associated substances potentially present in a leaking flow or by substances released in the environment following acidification (metals, organic compounds). Nevertheless, no information is available about this last point in the CCS field.

Pump-and-Treat

The pump-and-treat technique aims at remediating the contaminated section of an aquifer by extracting the contaminated water before treating it at the surface. It is often considered as one of the most common aquifer clean-up technologies for organic compounds, VOCs and dissolved metallic compounds (Bear and Sun, 1998 and US EPA, 1997). It can be designed either to restore the quality of aquifer or to limit the pollution by hydraulically containing the contaminants plume (Colombano et al., 2010, US EPA, 1997 and Bayer et al., 2002). The efficiency of this measure depends mainly on the aquifer properties and the pollutants considered (Khan et al., 2004); thus, cost and intervention delay are highly site specific (Colombano et al., 2010). According to Benson and Hepple (2005), this technology might be used when dissolved CO_2 or other contaminants (associated substances such as mobilized metals and organic compounds) are present in groundwater. CO_2 could be removed from pumped water by aeration.

Permeable Reactive Barrier (Treatment Wall)

The permeable reactive barrier aims at trapping or transforming contaminants from groundwater by physical, chemical or biological processes when groundwater flows across a treatment wall material set up for these purposes (US EPA, 1996). It is a rather recent technique but has already been applied in a relatively large number of sites and used generally to treat organic pollutants and metal contamination of groundwater (Colombano et al., 2010; Kahn et al., 2004). In case of CO_2 geological storage impacts remediation, this technique has been quoted by Benson and Hepple (2005) as a potential technique for the removal of mobilized trace elements. However, this technology is rather limited to shallow depths (Vidic and Pohland, 1996), which reduces its applicability to potential impacts of CO_2 geological storage.

Injection–Extraction

Different options based on fluid management in case of an accumulation of gaseous CO_2 in groundwater are suggested by Esposito and Benson (2012). These measures rely on extraction and injection techniques as well as combinations of both. Their aim is to extract the mobile/dissolved gaseous CO_2 or decrease the quantity of mobile CO_2 through residual trapping in the groundwater aquifer. Dilution and dispersion of the dissolved CO_2 might also be increased in case of water injection. A particular attention should be given to the exsolution of CO_2 that could occur due to pressure decrease during extraction. The efficiency of these techniques is highly dependent on leaking plume distribution; groundwater aquifer properties (permeability, anisotropy or presence of heterogeneities); injection–extraction configuration (number of wells, location and spacing or injection and extraction rates). The potential operational issues linked with fluid management corrective measures are further discussed in Réveillère et al. (2012). The measure is theoretically applicable. However, several issues need to be tackled in the case of the remediation of CO_2 migration in a groundwater aquifer: pressure increase (hydrodynamic impact), water availability or disposal, modification of water quality due to injection, and precise knowledge of the plume location.

In Situ Remediation with Microbes

Bacteria have the potential to biologically induce CO_2 mineralization into solid carbonate phases by a reaction called bioalkalinization, an ubiquitous phenomenon possible with various bacteria metabolisms. In addition to mineralization of CO_2, the induced pH increase could help counterbalancing the acidification provoked by CO_2 injection (Dupraz et al., 2009). In contrast with the addition of a base solution, which causes localized precipitation at injection point, the gradual increase of pH due to bacteria could provoke a wider spreading of the precipitation (Ferris et al., 2004). Moreover, Mitchell and Ferris (2005) suggested that the biologically induced co-precipitation of contaminants (e.g. heavy metals) with calcite precipitates could be a long term remediation option for contaminated groundwater. Few data are available about the cost and applicability at large scale of techniques relying on bacteria in the field of carbon storage remediation since results are essentially based on small scale (batch) laboratory experiments (Dupraz et al., 2009, Ménez et al., 2007 and Mitchell et al., 2010).

Remediation Techniques for Impacts in the Unsaturated Zone

The unsaturated zone is considered as the portion of the subsurface situated above the groundwater table. Its porosity is filled with air and water. Extensive measures and experience are available to remediate pollutions on the unsaturated zone. Possible impacts on the unsaturated zone in case of unexpected behaviour of a geological storage include lowering of soils pH and associated impacts, accumulation of gaseous CO_2 (and potentially associated substances) leading to asphyxiation of associated biota, leaching or mobilization of heavy metals or organic, and changes in bio-geo-chemical processes occurring in soils (Benson et al., 2002, IPPC, 2005, IEA-GHG, 2007b and US EPA, 2008). This could have subsequent impacts such as damage on surface ecosystems, and damage on economic activities relying on soil such as forestry and agriculture.

Monitored Natural Attenuation

Natural attenuation (see general description above) is also commonly used in environmental clean-up of soils (US EPA, 1999). As shown by Oldenburg and Unger (2003) and Zhang et al. (2004) through modelling, the unsaturated zone has a potential to naturally attenuate CO_2 leakages because of its buffer effect. The reduction of CO_2 concentration, and thus of the exposure level, is due to migration and dispersion/dilution in the atmosphere, in the porosity of soil layers and in the saturated zone. However, it may take 10 years or more to remediate a CO_2 plume (Zhang et al., 2004). Such a slow process (compared to active techniques) seems more acceptable for CO_2 due to its low toxicity compared to other pollutants (e.g. VOCs). Barometric pumping (natural flow of air due to natural cyclic variation of atmospheric pressure) reinforces the removal rate of CO_2 due to a larger flux to the atmosphere. In addition, natural processes in soil could reduce mobility or bioavailability of toxic compounds or transform them to another less toxic product. A major concern about natural attenuation is that the efficiency is extremely dependent on site specific conditions and is not straightforward to forecast.

Soil Vapour Extraction (SVE)

SVE, also known as soil venting or vacuum extraction, is an in situ treatment based on the establishment of a low pressure gradient in the unsaturated zone in order to force gases to flow towards extraction wells (US EPA, 2004 and Zhang et al., 2004). Once extracted from soil, the vapour phase is generally treated if direct emission in the atmosphere is not planned and/or not allowed. Detailed screening flow charts to evaluate whether or not SVE could be an appropriate remediation technique are available in the soil clean-up literature; see for example US EPA (2004). Particularly, the volatility of the constituents should be higher than 0.5 mm Hg (about 66 Pa): that is typically the case for CO_2. Though SVE has proven its effectiveness on numerous polluted sites for the last 30 years, it is still not obvious how to determine whether this technology will be efficient at a specific site (US EPA, 2004). This depends on many parameters such as: intrinsic permeability, soil liquid saturation, heterogeneities, thickness of the unsaturated zone, and depth of the water table (Poulsen et al., 1999, US EPA, 2004 and Khan et al., 2004). Treatment duration for SVE projects in appropriate conditions usually varies from 6 months to 2 years (Barnes, 2003 and Khan et al., 2004). Extensive design information and guidance documents to assist the SVE practitioner are available in soil clean-up field (e.g. US Army Corps of Engineers, 2002, US EPA, 1997 and US EPA, 2004). This remediation technique has been suggested by several authors as applicable to CO_2 storage sites (Benson and Hepple, 2005, Rohmer et al., 2010, Zhang et al., 2004 and Sweatman et al., 2010). Numerical simulations from Zhang et al. (2004) and Rohmer et al. (2010) showed that SVE could be, under certain conditions, an effective way to remove CO_2 from the unsaturated zone.

Depending on the situation, some options or modifications could enhance or be more appropriate than a basic SVE system, among which:

- Air sparging – this expands the remediation possibilities of SVE to the saturated zone. The method consists of injecting gas in the saturated zone below or within the polluted zone to enable phase transfer from a dissolved state to a vapour phase US EPA (2004). The gases are then vented through the unsaturated zone where they can be collected and treated. Air sparging has been used for several decades to reduce concentrations of VOCs below the water table (Khan et al., 2004). It could also theoretically be used in cases of dissolved

CO_2 due to analogies between CO_2 and VOCs (De Lary and Rohmer, 2010);

- Setting of an impermeable surface cover – it decreases or stops the flux at the soil surface and prevents short circuiting by the air of the atmosphere during pumping (Benson and Hepple, 2005 and Zhang et al., 2004);
- Directional drilling (Zhang et al., 2004 and US EPA, 1997);
- Pumping of CO_2 in horizontal trenches dug on purpose where it accumulates due to its density (Benson and Hepple, 2005);
- Injection and passive inlet wells – they could be used to enhance the transport of CO_2 in the unsaturated zone (US EPA, 2004) by injection or passive introduction of air in soil.

pH Adjustment

The objective of pH adjustment is to increase the pH to neutralize soil acidification consequences (damage to ecosystem or crops/forest) or to prevent possible induced impacts (such as leaching of heavy metals due to acidification). Adjustment of pH has been carried out for a long time in agriculture (Ristow et al., 2010). Benson and Hepple (2005) and Sweatman et al. (2010) proposed alkaline supplements (lime) spreading to remediate acidification of soil due to a potential leakage of CO_2 in soil. Irrigation and drainage of soil might be another way to adjust the pH of soil by dilution of CO_2 into groundwater and/or by using a pump-and-treat system once CO_2 is dissolved (Benson and Hepple, 2005).

Remediation Techniques for Impacts on Surface Assets

Surface Water

Following an undesired migration from the reservoir, CO_2 (or other associated contaminants) may dissolve in rivers, ponds or lakes, where it could accumulate. If the CO_2 flux exceeds the dissolution capacity of the water, then CO_2 bubbles will form and part of the leaking flux will quickly transfer to the atmosphere. FromBenson and Hepple (2005), if the water body is shallow and/or well-mixed (shallow lake or pond) or turbulent (streams or rivers) CO_2 will be released into the atmosphere where it will disperse quickly. In this situation, monitored natural remediation could be an appropriate measure. Although it seems very unlikely to occur, accumulation in deep stratified lake due to leakage from CO_2 storage could be remediated though a system for degassing set up similar to that on Lake Nyos and Lake Monoun: a vertical pipe between the lake bottom and the surface provokes a permanent controlled eruption of CO_2 (Benson and Hepple, 2005 and Halbwachs et al., 2004).

Indoor Environment

To manage possible chronic intrusion of gaseous CO_2 into buildings, Benson and Hepple (2005) andRohmer et al. (2010) suggested the use of techniques that have been developed to remediate radon or organic compounds that have intruded into buildings or to prevent any intrusion including (Benson and Hepple, 2005, IDELG, 2002 and US EPA, 2001): sealing the openings in the building, sub-floor (sub-slab) depressurization with passive system or electrical fans, sub-floor pressurization to force soil gases to migrate away from the building; adjustment of ventilation and of indoor pressure. There are few field data and experiments about remediation of CO_2 intrusion into buildings (Robinson, 2010), meaning that the maturity level of such measures in this field is still low.

Atmosphere

Unexpected CO_2 migration in the subsurface could potentially lead to gas releases to the atmosphere. At natural sites with important soil CO_2 concentrations and high flux at surface, the exposure concentration at one or two metres above ground is low because CO_2 disperses rapidly in the open atmosphere (e.g.,Carapezza et al., 2003 and Farrar et al., 1999). However, dispersion (dilution) is strongly dependent on the meteorological and topographical conditions (Chow et al., 2009 and Oldenburg and Unger, 2004). Some active measures can be set up in low lying areas and for the cases when natural mixing is not enough to disperse the CO_2 plume, e.g. air jets, helicopters or large fans (Koornneef et al., 2012 and Sweatman et al., 2010). At large scale, natural atmospheric mixing would be the only practical method to lead to the natural dilution of CO_2 leakages (Benson and Hepple, 2005).

Ecosystem Restoration

Numerous examples of successful ecosystem restoration projects already exist around the world; nevertheless it is still an area of on-going research (Benayas et al., 2009 and Nellemann and Corcoran, 2010). Depending on the impacts, the site specific conditions and the relevant regulations, a wide range of actions could be performed: restoring of vegetation or habitat, environment clean-up, reintroduction of species, etc. Natural recovery (recovery based on natural capacity without direct intervention) is an option to be considered in the ecosystem restoration process. Generic guidelines for restoration project managers and policy maker have been established (e.g. Clewell et al., 2004). More information is necessary about potential impacts on ecosystem of CO_2 leakages to be able to suggest, for CCS, some appropriate restoration measures.

Table 3 summarizes the measures suggested in the literature for remediation of impacts following a CO_2 leakage.

Table 3: Summary of proposed measures for impact remediation in the field of CO_2 storage

Impacted compartment	Suggested measure	Possible application in CCS domain
Groundwater	Monitored natural attenuation	- Reduction of contaminants concentration: e.g. aqueous CO_2 concentration (Benson and Hepple, 2005), associated substances such as mobilized metals and organic compounds. - Transformation of contaminants into less toxic products: e.g. associated substances such as metals, organic compounds. - Reduction of constituent mobility and bioavailability: e.g. associated substances such as metals, organic compounds.
	Pump-and-treat	- Extraction and treatment of fluids containing dissolved CO_2 or other contaminants (associated substances such as mobilized metals, organic compounds) (Benson and Hepple, 2005).
	Air sparging	- Volatilization and extraction of dissolved CO_2 and additional contaminants (with properties similar to VOCs) (De Lary and Rohmer, 2010 and Rohmer et al., 2010).
	Permeable reactive barrier (treatment wall)	- Trapping through a permeable barrier favouring reactions of mobilized trace elements (associated substances such as metals, organic compounds) (Benson and Hepple, 2005).
	Injection-extraction	- Extraction of the mobile gaseous plume. - Decrease of the quantity of mobile CO_2 in the groundwater aquifer. - Extracting the dissolved CO_2 and potential additional contaminants (Esposito and Benson, 2012).
	Remediation using microbes	- Adjustment of ground water pH (Dupraz et al., 2009). - Mineralization of dissolved CO_2 (Ménez et al., 2007). - Co-precipitation of contaminant (heavy metals) (Mitchell and Ferris, 2005).

Unsaturated zone	Monitored natural attenuation	- Reduction of CO_2 concentration in soil (Benson and Hepple, 2005, Sweatman et al., 2010 and Zhang et al., 2004). - Transformation or reduction of mobility of contaminants (e.g. organic compound, heavy metals).
	Soil vapour extraction	- Extraction of CO_2 (or organic compounds) from soil (Benson and Hepple, 2005, Rohmer et al., 2010, Zhang et al., 2004 and Sweatman et al., 2010).
	pH adjustment (spreading of alkaline supplements, irrigation and drainage)	- Adjustment of soil pH (Benson and Hepple, 2005 and Sweatman et al., 2010).
Surface water	Passive systems: natural attenuation	- Reduction of CO_2 concentration in shallow water (Benson and Hepple, 2005)
	Active venting system	- Removal of dissolved CO_2 in deep stratified lakes (Benson and Hepple, 2005)
Indoor environment	Usual remediation techniques (radon, VOC, etc.): sealing the opening, (de)pressurization, adjustment of ventilation	- Lowering of CO_2 concentrations in indoor air (Benson and Hepple, 2005 and Rohmer et al., 2010).
Atmosphere	Passive system: natural mixing	- Reduction of CO_2 exposure in the atmosphere (Benson and Hepple, 2005 and Sweatman et al., 2010).
	Air jets, helicopters or large fans	- Reduction of CO_2 exposure in the atmosphere (Benson and Hepple, 2005, Koornneef et al., 2012 and Sweatman et al., 2010).
Ecosystems	Ecological restoration	- Restoration of impacted ecosystem (if needed).

EXPERIENCE GAINED FROM PREVIOUS AND ONGOING CO$_2$ STORAGE PROJECTS

CO$_2$ geological storage has now been implemented in sites around the world[2]. In such projects, risk management procedures have been set up and mitigation/remediation techniques have been integrated into these procedures in case of deviation from the expected behaviour of the storage complex. Several guidelines or guidance documents mentioning the mitigation and remediation measures plan have been issued (EC, 2011, IEA, 2010, DNV, 2009 and DNV, 2012). A Canadian standard (CSA Z741) establishes requirements and recommendations for CO$_2$ geological storage, including one section on risk treatment (CSA, 2012). International standards (ISO 31000), which are not specific to CO$_2$ geological storage, also propose generic workflows for risk management and guidelines for risk treatment (ISO, 2009). However, a widely accepted methodology for designing intervention and remediation plans for CO$_2$ geological projects is missing. Moreover, to date, no comparison has been published between the different intervention strategies elaborated at existing storage sites or storage projects. Therefore, there is a need for establishing the best field-applied and tested practices for mitigating an undesired CO$_2$ migration, based on the information available in the literature, personal knowledge and expertise of the authors, and on the experience gained in the existing CO$_2$ storage projects.

Survey of Project Operators

Fourteen CO$_2$ storage or natural gas storage operating companies were contacted in order to gather information on their corrective measures plans. Two publicly available corrective measure or risk management plans have also been considered (the Goldeneye project; ScottishPower CCS Consortium, 2011 and the Gorgon project; Chevron, 2005 and Chevron, 2008). Among the 14 companies contacted, 8 participated in the survey through a written response or an oral interview. Fig. 1 shows the location, the type of project operated by the respondent and its status.

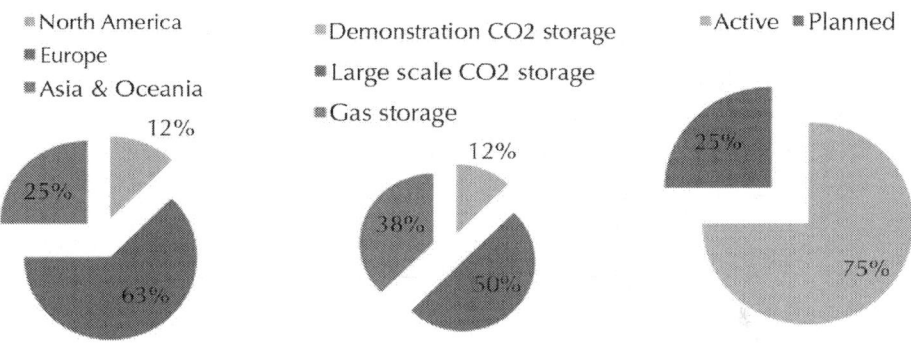

Figure 1: Breakdown of the answers per geographical origin, activity type, and status.

For comprehensiveness, this review also integrates the existing regulatory documents specific to CO_2 geological storage mentioning intervention plans: the European Directive (EC, 2009) and associated guidance document 2 (EC, 2011); the USA Federal Requirements (US EPA, 2010a); the Australian Act (Australian Government, 2011) and regulations (Australian Government, 2012). The Carbon Capture and Storage Model regulatory framework document (IEA, 2010), which addresses the regulatory issues associated to CCS and the Canadian Standard on geological storage of CO_2 (CSA, 2012) are also included.

Mitigation and Remediation Plans

Risk Based Corrective Measures Plan

According to the responses, the corrective measures plan appears to be site and project specific and based on the risk assessment outcomes. Risk identification is typically assessed through experts workshops using e.g. FEP methodology and/or bow-tie analyses in order to identify relevant risk scenarios. Typically, these workshops include mostly experienced people from oil and gas industry, and the risks are assessed in the context of their knowledge and experience.

Classically, for risk scenarios considered as critical after assessment, both monitoring and mitigation (or remediation) measures are contemplated in the corrective measures plan. Even if this review is exclusively focused on mitigation and remediation technologies, the importance of monitoring within the mitigation strategy has been emphasized in many responses.

In some projects, the corrective and/or remediation measures are implicitly included as risk reduction measures in the risk management plan, whereas in other projects they are presented in a distinct document. The European Directive (EC, 2009) for instance, requires a specific corrective measures plan for every storage permit application, which should however be linked with the risk assessment and risk monitoring plan. The EC Directive guidance document 2 (EC, 2011) proposes that a section of the corrective measure plan (*N.B.* Section 1) should be devoted to this link by including the following elements:

- The identified risk the measure is related to;
- The threshold triggering the corrective measure implementation;
- The monitoring system necessary for monitoring the measure effectiveness.

The Canadian Standard on geological storage of CO_2 (CSA, 2012) also recommends a risk treatment plan for each significant risk, with the following elements:

- The objective in terms of target level of risk to be achieved;
- A prioritized list of preferred risk treatment options;
- The analysis to be performed to ensure an acceptable level of risk over time;

- The cyclic assessment of the effect of the implemented risk treatment options and the acceptability of the residual risk level (followed by new risk treatment options if it is assessed as non-acceptable);

- A contingency plan for unexpected circumstances or incidents.

In the Gorgon project (Chevron, 2005), the corrective measures are implicitly included in the risk management plan. A CO_2 injection management plan has first been set up, applying the practice of reservoir management plans developed in the oil and gas industry. "Management actions" i.e. strategies and methods are proposed in order to mitigate several situations. Mitigation measures are also recapped in the second phase of the analysis, called the potential failure modes assessment, which includes the failure modes resulting in the unplanned migration of CO_2. The Goldeneye project (ScottishPower CCS Consortium, 2011) explicitly sets up a corrective measures plan in the format required by the European Commission and follows the guidelines provided by the Commission for the plan elaboration (EC, 2011). The potential subsurface migration paths identified in the risk assessment plan are recalled and served as a basis for the proposed corrective measures.

Reviewed and Updated Corrective Measures Plans

All the respondents to the survey mentioned that their corrective measures plan had been reviewed. The review was done by one or several of the following: the stakeholders who had significant exposure in the project, the IEA-GHG, the owner or representatives of the operator, engineering companies, technical teams (including experts in the field of geological CO_2 storage), HSE department, fire-fighting services in some cases and regulatory authorities. Several respondents have also mentioned that their plan had been updated after this review, and/or that it is planned to be regularly updated. Plans are also updated if new information comes up during operations from injection and monitoring which may change the risk profile. An example of such an update is provided from the Gorgon project in Australia, which adapted the initial failure modes assessment and the associated mitigation measures to the proposed increase in injection rates (Chevron, 2008). The USA Federal Requirements (US EPA, 2010a) impose such updates after a minimum fixed frequency (not to exceed five years), following any significant change, or when required by the Competent Authority. No regular update requirement of the corrective measures plan is mentioned in the EC Directive (EC, 2009); the storage permit review might however impose such updates in case of new information regarding monitoring, site characterization, risk assessment or scientific knowledge (including new corrective measures or methods) (EC, 2011). According to the Australian regulations (Australian Government, 2012), well operations management plans should be modified in case of changes in the understanding about the characteristics of geology, in case of the occurrence of a new risk or of a significant increase in a detrimental risk. The responses of operators of large scale CO_2 storage sites highlighted the fact that their corrective measures plan or summaries were in the public domain, either in reports or articles. In addition to public plans or summaries, internal versions contain additional specifications as well as cost information. One response insisted on the necessity of obtaining commitment from all diverse and required parties, which might be challenging given the different concerns and drivers.

Mitigation and Remediation Methods and Techniques Included in Plans

The statements obtained from the survey responses on the methods and techniques included in plans corroborate the EC Directive guidance document (2011) that points out the limited experience in terms of corrective measures in the CCS field and the reliance on existing rules and regulations developed by and established in the oil and gas industry. The methods need to be adapted mostly from oil and gas and environmental clean-up industries.

The CCS Model regulatory framework document (IEA, 2010) also confirms that best practices already exist in the oil and gas industry regarding mitigation measures such as well-plugging or well-repair techniques. Partial removal of CO_2 from a reservoir, decreasing the reservoir pressure and remediation of groundwater in case of impacts are also mentioned.

The survey also pointed out the lack of precise knowledge of the processes that might be occurring during an unexpected CO_2 migration. In parallel to the development of mitigation and remediation techniques, the behaviour of leakage over time should be better understood to be able to choose the most relevant mitigation and remediation strategies.

In the majority of the reviewed corrective measures plans for CO_2 storage sites, the measures were classified depending on the type of migration pathways: either natural (e.g. fractures, faults) or man-made (i.e. related to wells). This distinction is essential when dealing with risk treatment as the effectiveness is likely to be much more limited for the geological system, as observed in the EC Directive guidance document (EC, 2011). In the remainder of this section, responses received relating to the inclusion of the measures presented in section 2 are stated.

Action on Wells

In all risk assessments, leakages through operating or abandoned wells have been identified as potential risks. All CO_2 storage projects are concerned since they have at least one operating well used for injection, and possibly more abandoned or operating wells, usually related to oil and gas exploration and/or production. Intervention on wells is a large part of the corrective measures considered in all reviewed plans. It is based on oil and gas industry experience and many different techniques are proposed: well killing; injection tubing and packer replacement, repair (scab casing, squeeze cementing, patch or sealant) or plugging and abandonment.

These interventions are specific to each well. The measures potentially applied to the injection well may be detailed: for instance, the remediation plan for Goldeneye project (ScottishPower Consortium, 2011) recalls the injection well completion, the identified potential leakage pathways and proposes corresponding intervention measures. However, the well remediation techniques are not always specified in the plans, especially regarding abandoned wells whose completion is less likely to be precisely known, but such techniques are known to be available and applicable based on oil and gas industry experience.

CO_2 blowout risks were mentioned and one project even mentioned a written blowout contingency plan associated to some CO_2 release modelling based on a major leak at the wellhead.

Fluid and Pressure Management

Fluid management options are primarily considered in case of geological leakage pathways, unpredicted CO_2 plume migration, or unacceptably high pressure build-up in the near injection well formation. The measures considered include turning the injection off, reducing the injection flow rate or varying the injection pattern. In addition, extending vertically the overall perforation interval or perforating new reservoir zones for CO_2 injection were two fluid management options cited several times in the responses. One operator has incorporated the drilling of a new well for CO_2 injection into its mitigation plan aiming at minimizing operational risks. Once the new injector is in place, the plans are that the currently existing CO_2 injector will be shut down and used as a backup in case of an operational down-time of the new CO_2 injector or used for pressure monitoring.

However, most of the large scale CCS projects answered that no pressure relief well was considered, mentioning that it was either not pertinent (e.g. in case of storage in a depressurized depleted reservoir) or extremely expensive especially in offshore locations. Other projects considered the drilling of a new injection or pressure relief well as an ultimate measure. No project considered producing back the injected CO_2 from the injection well or locally in the vicinity of an identified leakage. No project considered injecting brine or water for enhancing residual trapping and dissolution, or implementing a hydraulic barrier (which has been judged conceivable but not achievable at an affordable cost). Pressure management strategies are rather seen as preventive measures as, for instance, the drilling of four extraction wells in addition to the injection wells at the Gorgon project. The extracted water is proposed to be injected into an overlying geologic unit (Chevron, 2008). Modelling studies using planned injection and extraction wells have been undertaken (IEA-GHG, 2012) and show that the combination of these wells is beneficial both for pressure management and plume migration control.

The responses received are in line with the EC Directive guidance document (EC, 2011), which states that whereas some fluid management procedures are quite classical, notably in the oil and gas industry, others are either novel techniques never applied in situ or those that are technically feasible but only at very high cost.

Breakthrough Technologies

Breakthrough technologies such as foams, gels or other low-permeability materials were not explicitly mentioned in any of the responses to the survey. The main reasons were that some of these technologies are either not mature enough or not tested extensively enough for the size/type of the problem which may be encountered in CO_2 storage sites. If new methods are developed and tested after the establishment of a mitigation measures plan, these methods should be added during the periodic plan update. Moreover, an intervention measure will ultimately be decided at the time of the leakage based on the available technologies, which may, in the future, include these nowadays-breakthrough technologies.

Remediation of Impacts in Sensitive Aquifers, in the Vadose Zone or at Surface

The risk assessments which form the basis of the reviewed

corrective measures plans, concluded that the risks of impacts in sensitive aquifers, in the vadose zone or at the surface were low either because existing natural barriers (e.g. secondary seals) make it unlikely that leaked CO_2 would reach vulnerable assets or because vulnerable assets such as vadose zone or drinking water aquifers were not present (e.g. in offshore sites). The development in advance of customized remediation measures of impacts was often judged unnecessary. The low risk of contaminations relevant to the significant experience in the environmental clean-up industry explains the fact that no remediation method for these types of impacts was presented in the reviewed CCS projects.

Flexible Decision Protocol and Holistic Decision Making

Flexible Protocol for Measure Implementation

Most answers clearly stated that corrective measures plans are submitted by the operator during the storage permit application. These plans are based on identified leakage scenarios established during the risk assessment process. However, the operator and the Competent Authority do not know the specific location and the actual process of a significant irregularity or leakage before it is detected. The best measure design in response to the leakage may not be submitted a priori in the plan. Even if it states that plans should be "ready to use", the EC Directive guidance 2 document (EC, 2011) also acknowledges the fact that the corrective measures plan might be generic, especially at the first stages of the storage site lifecycle.

The corrective measures plan should therefore allow for flexibility and, ultimately, the final decision and measure design will be taken by the operator and the Competent Authority at the time of leakage. DNV recommended practices (DNV, 2012) state that the implementation of risk treatment should be done according to the following procedure: (1) *the detection of a circumstance that signals the need to implement risk treatment*; (2) *assessment and selection of an appropriate treatment to address the situation*; (3) *the implementation of the selected risk treatment*. This should be followed by a risk assessment step to assess whether an additional treatment action should be carried out (DNV, 2012). Several plans include this classical decision process for implementing corrective measures. The Federal Requirements under the Underground Injection Control Program for CO_2 Geologic Sequestration Wells (US EPA, 2010a) provides as well a detailed procedure to be followed in case of a potential impact on groundwater: cease injection, identify and characterize any release, notify to the competent authority, implement the emergency and remedial response approved by the competent authority.

In terms of responsibility, the IEA CCS model regulatory framework document (IEA, 2010) explains that the existing regulatory documents tend to give the responsibility of implementing the measures to the operator, while the relevant authority would ultimately decide which measures should be considered and whether this implementation is necessary or not.

Holistic Approach during the Decision Process

One of the responding operators insisted on the fact that, after detecting and analysing the irregularity, the team in charge of deciding and implementing the most appropriate action should have a systematic approach considering all possible corrective measures given the type of risk to mitigate, their economic cost, operational feasibility, and potential environmental impacts with and without implementing the measure. That team should aim at an overall risk reduction; one option may be not to intervene if the abnormal migration does not threaten a sensitive target or if the risks associated with the measure itself are too high compared to its benefits. The decision should be made having a holistic view, balancing the risk of the leakage, and the practically achievable options. According to DNV (2012), risk treatment options should be identified among the methods that are cost effective and do not introduce other significant risks which outweigh potential benefits of the treatment. In order to allow comparability between relevant measures, the EC Directive guidance 2 document (EC, 2011) proposes a possible format for the corrective measures plan. This format notably specifies details about the measures likely to be needed for decision making: *estimated timeframe needed for implementation, detailed description of the measure* (including activities to be carried out), *rationale for the use of the measure, current status of the measure* (proven, commercial, under development).

CONCLUDING REMARKS: BEST PRACTICES AND FUTURE CHALLENGES

The development of CO_2 capture and geological storage technology is highly dependent on the assurance of the storage process safety, especially with regard to a potential CO_2 leakage out of the target zone. A proper risk management process should be set up for this aim and integrated from the very start of the project because site selection and safe operation are the most crucial preventive measures. The process will include assessing the risks specific to a given storage site, monitoring the site to detect any potential loss of confinement, mitigating a potential leakage and remediating possible impacts. As outcomes of this study, some recommendations are proposed and some research directions are suggested for filling the existing gaps to mitigate unpredicted CO_2 migration.

Best Practices Stemming from the Review

The intervention plan should, at a given date, mention the technologies available according to the state of knowledge. The selection criteria used for a knowledgeable choice of the measures should be also part of this plan. The plan should be reviewed and updated to allow the integration of any new measures or any new information that may change the ranking list of measures or the associated information.

The intervention plan should be fully integrated in the global risk management process. The mitigation and remedial actions should therefore be linked with the risk scenario selected during the risk assessment process and each measure should be related to the irregularities it mitigates or to the effect it remediates. In addition, the methods described in the monitoring plan should be mentioned in the intervention plan. Monitoring is essential to quickly detect irregularities and give appropriate information for implementation of measures when needed. Equally important is the ability of the monitoring scheme to control the efficiency of any implemented mitigating measure and assess needs for corrections to the intervention plans.

The intervention plan should be flexible since the choice and design of the measure is dependent on the situation: the plan should be somewhat generic proposing adapted measures to a potential situation; however, it should be specific enough to help as much as possible the decision-making if a deficiency occurs. Therefore, it should also include tools to allow the choice and design of the most relevant mitigation and remediation strategy. The final decision will be ultimately made taking into account the specifics of the situation and considerations/approaches of both operators and regulators. Nevertheless, general principles, based on common practices in other fields, could be followed such as the implementation of a set of independent measures rather than of only one measure, the preferential use of techniques removing the risk source or of passive measures especially after the site closure.

Future Challenges

Gaps in Knowledge and Technologies

Most of the measures are originated from the experience in mitigation or remediation of other kinds of risks and impacts (oil and gas industry and environmental clean-up). Even if the analogy is meaningful, CO_2 geological storage brings new conditions and requirements. Therefore, there is a need for researching how these measures could be adapted to the specific conditions of CO_2 geologic storage. For instance, remediation measures originated from the environmental clean-up field are very much referenced, but their applicability to CO_2 migration potential impacts have been hardly studied, contrary to technologies from the oil and gas sector, which are the basis of the measures proposed in the literature or submitted in existing remediation plans. The development of new measures tailored to CO_2 leakage is also needed. Some theoretical concepts have been proposed for instance in terms of pressure management; however, their feasibility has to be proven through experimental tests, in situ deployment, or experimentations at a scale between the lab and the field following a detailed modelling and simulation studies. Similarly, as presented in this study, some breakthrough technologies are being developed; however, their development is at early stages and therefore much effort is needed to integrate them in the portfolio of mitigation and remediation measures. In particular, these efforts could be focused on methods to modify the properties of natural pathways (such as faults or relatively more permeable zones in the caprock), as few solutions seem to be available to deal with this issue.

In practice, the success of an intervention will be highly dependent on the knowledge of what is actually happening at the storage site. This implies knowing location and nature of the irregularity or impact to be treated: at present, many leakage phenomena are not well understood (e.g. the mechanisms underlying the migration of CO_2 across several geological formations), and therefore the potential impacts cannot be well characterized. A deeper understanding of the mechanisms of migration of CO_2 can also improve interpretation of monitoring data. Research on migration processes should therefore be pursued.

Operational Needs

As well as a list of measures to be potentially applied in case of unwanted migration of CO_2, the operators or regulators will need a comprehensive description of each measure in order to make an informed choice. The purpose of the measure, the time needed for implementation, the associated economic costs, the maturity or the environmental impacts of a measure are key factors that need to be assessed. In general, there is a lack of such information and therefore an extensive work is needed to fill this gap.

One of the main challenges related to mitigation and remediation of leakage in the field of CO_2 storage is to choose the best possible way to intervene, and to do so at a reasonable cost. Moreover, a negotiation may take place between operators and regulators regarding a specific event because they do not share the same objectives. A method, based on multi-criteria analysis (and/or cost-benefit analysis) and adapted to the context of mitigation and remediation of irregularities in CO_2 storage could help to support informed, shareable and more acceptable decisions. The development of such method might require the creation of specific approaches for assessing a priori the effectiveness of a measure. In addition, to ensure robustness and acceptability of this type of analysis, shared best practices are needed.

ACKNOWLEDGMENT

This work has been funded primarily by the IEA Greenhouse Gas R&D Programme and performed by BRGM and IRIS, members of CO2GeoNet, the European Network of Excellence on CO_2 geological storage (www.co2geonet.eu). A comprehensive version

of this review can be found in the report IEA/CON/12/203 –
Methodologies and technologies for mitigation of undesired CO_2
migration in the subsurface. The authors would like to thank the
industrial companies for their kind responses to the survey and
the anonymous reviewers for their comments, which, respectively,
contributed to the technical contents and helped to improve the
published version of this article.

REFERENCES

1.   Abshire, L., Desai, P., Mueller, D., Paulsen, W.B., Robertson,
      R.D.B., Solheim, T., 2012. Offshore permanent well
      abandonment. Oilfield Review 24, 42–50 (Spring).

2.   Abshire, L., Hekelaar, S., Desai, P., 2013. Offshore plug &
      abandonment: challenges and technical solutions. In: Paper
      OTC 23906 Presented at the Offshore Technology Conference,
      Houston, TX, USA, May 6–9.

3.   Akervoll, I., Lindeberg, E., Lackner, A., 2009. Feasibility of
      reproduction of stored CO2 from the Utsira formation at the
      Sleipner gas field. Energy Procedia 1 (1), 2557–2564.

4.   André, L., Audigane, P., Azaroual, M., Menjoz, A., 2007.
      Numerical modeling of fluid–rock chemical interactions at the
      supercritical CO2–liquid interface during CO2 injection into
      a carbonate reservoir, the Dogger aquifer (Paris Basin, France).
      Energy Conversion and Management 48 (6), 1782–1797.

5.   Australian Government, 2011. Offshore Petroleum and
      Greenhouse Gas Storage Act 2006.

6.   Australian Government, 2012. Offshore Petroleum and
      Greenhouse Gas Storage (Resource Management and
      Administration) Regulations 2011 made under the Offshore
      Petroleum and Greenhouse Gas Storage Act 2006.

7.   Barclay, I., Pellenbarg, J., Tettero, F., Pfeiffer, J., Slater, H., Staal,
      T., Stiles, D., Tilling, G., Whitney, C., 2002. The beginning
      of the end: a review of abandonment and decommissioning
      practices. Oilfield Review 13, 28–41.

8.   Barlet-Gouedard, V., Rimmele, G., Goffe, B., Porcherie, O.,2006. Mitigation strategies for the risk of CO2 migration through wellbores. In: IADC/SPE 98924 Drilling Conference. Society of Petroleum Engineers, Miami, FL, USA.

9.   Barnes, D.L., 2003. Estimation of operation time for soil vapour extraction systems. Journal of Environmental Engineering 129 (9), 873–878.

10.   Bayer, P., Finkel, M., Teutsch, G., 2002. Reliability of hydraulic performance and cost estimates of barrier-supported pump-and-treat systems in heterogeneous aquifers. IAHS-AISH Publication 277, 331–338.

11.   Bear, J., Sun, Y., 1998. Optimization of pump-treat-injection (PTI) design for the remediation of a contaminated aquifer: multi-stage design with chance constraints. Journal of Contaminant Hydrology 29, 225–244.

12.   Benayas, J., Newton, A., Diaz, A., 2009. Enhancement of biodiversity and ecosystem services by ecological restoration: a meta-analysis. Science 325, 1121–1124.

13.   Benson, S., Hepple, R., 2005. Prospects for early detection and options for remediation of leakage from CO2 storage projects. In: Thomas, D.C., Benson, S.M. (Eds.), Carbon Dioxide Capture for Storage in Deep Geologic Formations, vol. 2. Elsevier Ltd., Elsevier Publishing, UK (Chapter 28).

14.   Benson, S.M., Hepple, R., Apps, J., et al., 2002. Lessons Learned from Natural and Industrial Analogues for Storage of Carbon Dioxide in Deep Geologic Formations. Lawrence Berkeley National Laboratory Report, LBL-51170, Berkeley, CA.

15.   Bergmo, P.E.S., Alv-Arne Grimstad, A.-A., Lindeberg, E., 2011. Simultaneous CO2 injection and water production to optimise aquifer storage capacity. International Journal of Greenhouse Gas Control 5, 555–564, http://dx.doi.org/10.1016/j.ijggc.2010.09.002.

16.   Birkholzer, J.T., Cihan, A., Zhou, Q., 2012. Impact-driven pressure management via targeted brine extraction—

conceptual studies of CO2 storage in saline formations. International Journal of Greenhouse Gas Control 7, 168–180.

17. Burns, L.D., Burns, M., Wilhite, P., McCool, S., Oglesby, K., Glass, J, 2008. New generation silicate gel system for casing repairs and water shutoff. In: Paper SPE 113490 Presented at the SPE/DOE Symposium on Improved Oil Recovery, Tulsa, OK, 20–23 April.

18. Burton, M., Bryant, S.L., 2009. Eliminating buoyant migration of sequestered CO2 through surface dissolution:implementation costs and technical challenges. SPE Reservoir Evaluation & Engineering, 399–407.

19. Carapezza, M.L., Badalamenti, B., Cavarra, L., et al., 2003. Gas hazard assessment in a densely inhabited area of Colli Albani Volcano (Cava dei Selci, Roma). Journal of Volcanology and Geothermal Research 123 (1–2), 81–94, R&D Publication 95.

20. Carey, M.A., Finnamore, J.R., Morrey, M.J., et al., 2000. Guidance on the Assessment and Monitoring of Natural Attenuation of Contaminants in Groundwater. Environment Agency R&D Dissemination Centre, c/o WRc, Frankland Road, Swindon, Wilts SN5 8YF.

21. Carey, M.A., Wigand, M., Chipera, S., Gabriel, G.W., Pawar, R., Lichtner, P.C., Wehner, S.C., Raines, M.A., Guthrie Jr., G.D., 2007. Analysis and performance of oil well cement with 30 years of CO2 exposure from the SACROC unit, West Texas, USA. International Journal of Greenhouse Gas Control, 75–85.

22. Chevron, 2005. Greenhouse Gas Emissions – Risks and Management, Gorgon joint venturers, Available at: http://www.chevronaustralia.com/Libraries/

23. Chevron Documents/Rev O ch13 23Aug05.pdf.sflb.ashx (Chapter 13).

24. Chevron, 2008. Gorgon Gas Development Revised and Expanded Proposal – Public Environmental Review, Available at: http://www.chevronaustralia. com/ourbusinesses/gorgon/environmentalresponsibility/environmental approvals.aspx

25. Chow, F.K., Granvold, P.W., Oldenburg, C.M., 2009. Modeling the effects of topography and wind on atmospheric dispersion of CO2 surface leakage at geologic carbon sequestration sites. Energy Procedia 1, 1925–1932.

26. Chustz, M., Mason, D., Schober, J., De Lucia, F., Butterfield, C., Didyk, V., 2005. Expandable liner installation avoids sidetracking following production casing failure in the Gulf of Mexico. In: Paper SPE/IADC 91923 Presented at the SPE/IADC Drilling Conference, Amsterdam, Netherlands, 23–25 February.

27. Cirer, D., Arze, E., Moggia, J.M., Soto, H.R., 2012. Practical cementing technique to repair severe casing damage. In: Paper SPE/ICoTA154241 Presented atthe Coiled Tubing and Well Intervention Conference and Exhibition, The Woodlands, TX, USA, 27–28 March.

28. Clewell, A., Rieger, J., Munro, J., 2004. Guidelines for Developing and Managing Ecological Restoration Projects, 2nd ed. www.ser.org & Tucson: Society for Ecological Restoration International.

29. Colombano, S., Saada, A., Guerin, V., Bataillard, P., Bellenfant, S., Beranger, S., Blanc,

30. C., Zornig, C., 2010. Quelles techniques pour quels traitements – Analyse coûtsbénéfices, Rapport final. BRGM-RP-58609, 403 pp.

31. Crow, W., Carey, J.W., Gasda, S., Williams, D.B., Celia, M., 2010. Wellbore integrity analysis of a natural CO2 producer. International Journal of Greenhouse Gas Control 4 (2), 186–197.

32. CSA (Canadian Standards Association), 2012. CSA Z741 Geological storage of carbon dioxide.

33. Cunningham,A.B., Sharp,R.R.,Hiebert,R.,James, G., 2003. Subsurface biofilmbarriers for the containment and remediation of contaminated groundwater. Bioremediation Journal 7 (3–4), 151–164.

34. Cunningham,A.B., Gerlach,R., Spangler, L.,Mitchell,A.C.,

2009.Microbially enhanced geologic containment of sequestered supercritical CO2. GHGT-9. Energy Procedian 1, 3245–3252, http://dx.doi.org/10.1016/j.egypro.2009.02.109.

35.   Cunningham, A.B., Gerlach, R., Spangler, L., Mitchell, A.C., Parks, S., Phillips, A., 2011. Reducing the risk of wellbore leakage of CO2 using engineered biomineralization barriers. GHGT-10. Energy Procedia 4, 5178–5185, http://dx.doi.org/10. 1016/j.egypro.2011.02.495.

36.   Daigle, C., Campo, D.B., Naquin, C.J., Cardenas, R., Ring, L.M., York, P.L., 2000. Expandable Tubulars: field examples of application in well construction and remediation. In: Paper SPE 62958 Annual Technical Conference and Exhibition, Dallas, TA, USA, 1–4 October.

37.   Davidovitch, J., 2005. Geopolymer, Green Chemistry and Sustainable Development Solutions. World Congress Geopolymer, Institute of Geopolymer, Saint-Quentin, France.

38.   De Lary, L., Rohmer, J., 2010. Intrusion de CO2 dans un aquifère libre de surface: evaluation de l'impact et propositions de stratégies de réparation, BRGM/RP- 59070-FR, 86 pp., 53 ill., 1ann.

39.   Deremble, L., Loizzo, M., Huet, B., Lecampion, B., Quesada, D., 2010. Assessment of leakage pathways along a cemented annulus. In: Paper SPE 139693 Presented at the SPE International Conference on CO2 Capture, Storage, and Utilization, New Orleans, Louisiana, November 10–12.

40.   DiCarlo, D.A., Aminzadeh, B., Roberts, M., Chung, D.H., Bryant, S.L., Huh, C., 2011. Mobility control through spontaneous formation of nanoparticle stabilized emulsions. Geophysical Research Letters 38, L24404, http://dx.doi.org/10.1029/2011GL050147.

41.   Dillon, P.W., Billings, W.C., 1994. Well completion method and apparatus using a scab casing. United States Patent 5346077.

42.   DNV, 2009. CO2QUALSTORE – Guideline for Selection and Qualification of Sites and Projects for Geological Storage of

CO2. DNV Report No.: 2009-1425.

43. DNV, 2012. Geological Storage of Carbon Dioxide – Recommended practice DNV-RP-J203, Available at: http://www.dnv.com/news events/news/2012/ newcertificationframeworkforco2storage.asp

44. Doughty, C., 2007. Modeling geologic storage of carbon dioxide: comparison of non-hysteretic and hysteretic characteristic curves. Energy Conversion and

45. Management 48, 1768–1781.

46. Dupraz, S., Parmentier, M., Ménez, B., et al., 2009. Experimental and numerical modeling of bacterially induced pH increase and calcite precipitation in saline aquifers. Chemical Geology 265, 44–53.

47. Durst, D.G., Ruzic, N.,2009. Expandable tubulars facilitate improved well stimulation and well production. In: SPE 121147 Western Regional Meeting. Society of Petroleum Engineers, San Jose, CA.

48. EC (European Commission), 2009. Directive 2009/31/EC of the European Parliament and of the Council of 23 April 2009 on the geological storage of carbon dioxide and amending Council Directive 85/337/EEC, European Parliament and Council

49. Directives 2000/60/EC, 2001/80/EC, 2004/35/EC, 2006/12, EC, 2008/1/EC and Regulation (EC) No 1013/2006.

50. EC (European Commission), 2011. Implementation of Directive 2009/31/EC on the Geological Storage of Carbon Dioxide. Guidance Document 2. Characterisation of the Storage Complex, CO2 Stream Composition, Monitoring and Corrective Measures.

51. Eke, P.E., Naylor, M., Haszeldine, S., Curtis, A., 2011a. CO2 leakage prevention technologies. In: Paper SPE 145263 Presented at the SPE Offshore Europe Oil and Gas Conference and Exhibition, Aberdeen, UK, 6–8 September.

52. Eke, P.E., Naylor, M., Haszeldine, S., Curtis, A., 2011b. CO2/brine surface dissolution and injection: CO2 storage

enhancement. SPE Projects, Facilities & Construction, 41–53.

53. Esposito, A., Benson, S.M., 2012. Evaluation and development of options for remediation of CO2 leakage into groundwater aquifers from geologic carbon storage. International Journal of Greenhouse Gas Control 7, 62–73.

54. Farra, C.D., Neil, J.M., Howle, J.F., 1999. Magmatic carbon dioxide emissions at Mammoth Mountain, CA. U.S. Geological Survey, Water-Resources Investigations Report 98-4217, 34 pp.

55. Ferris, F.G., Phoenix, V., Fujita, Y., et al., 2004. Kinetics of calcite precipitation induced by ureolytic bacteria at 10 to 20 °C in artificial groundwater. Geochimica et Cosmochimica Acta 67 (8), 1701–1722.

56. Friedmann, S.J., 2007. Geological carbondioxide sequestration. Elements 3, 179–184. Gasda, S.E., Bachu, S., Celia, M.A., 2004. Spatial characterization of the location of potentially leaky wells penetrating a deep saline aquifer in a Mature Sedimentary Basin. Environmental Geology 46 (6), 707–720.

57. Gasda, S.E., Nordbotten, J.M., Celia, M.A., 2008. Determining effective wellbore permeability from a field pressure test: a numerical analysis of detection limits.

58. Environmental Geology 54, 1207–1215.

59. Halbwachs, M., Sabroux, J.C., Grangeon, J., et al., 2004. Degassing the "Killer Lakes" Nyos and Monoun, Cameroon. Eos 85 (July (30)), 281–285.

60. Hatzignatiou, D.G., Riis, F., Berenblyum, R., Hladik, V., Lojka, R., Francu, J., 2011. Screening and evaluation of a saline aquifer for CO2 storage: central Bohemian Basin, Czech Republic. International Journal of Greenhouse Gas Control 5, 1429–1442, http://dx.doi.org/10.1016/j.ijggc.2011.07.013. Hatzignatiou, D.G., Helleren, J., Stavland, A., 2014b. Core-Scale Numerical Modeling of Chemical Flow-Zonal Isolation. SPE Production & Operations, Paper accepted for publication.

61. Hatzignatiou, D.G., Stavland, A., Khairil, R.R., Wessel-

Berg., D., 2014a. Foam Core Flooding: An Experimental and Numerical Modeling Approach. J. of Pet. Sci. & Eng, Paper accepted for publication.

62. Haughton, D.B., Connell, P.L., 2006. Method to reduce milling and trip times during the recovery operations of permanent production packers. In: Paper IADC/SPE 99138 Presented at the Drilling Conference, Miami, FL, USA, 21–23 February.

63. Humez, P., Audigane, P., Lions, J., Chiaberge, C., Bellenfant, G., 2011. Modeling of CO2 leakage up through an abandoned well from deep saline aquifer to shallow fresh groundwaters. Transport in Porous Media 90, 153–181. Ide, T.S., Friedmann, S.J., Herzog, H.J., 2006. CO2 leakage through existing wells: current technology and regulations, GHGT-8. In: 8th International Conference on

64. Greenhouse Gas Control Technologies, Trondheim, Norway. IDELG (Irish Department ofthe Environment and Local Government), 2002. Radon in Existing Building. Corrective Options. Published by the Stationery Office, Dublin.

65. IEA (International Energy Agency), 2010. Carbon Capture and Storage – Model Regulatory Framework. Information paper.

66. IEA-GHG (IEA Greenhouse Gas R&D Program), 2007a. Remediation of leakage from CO2 storage reservoirs, Report 2007/11.

67. IEA-GHG (IEA Greenhouse Gas R&D Program), 2007b. Study of Potential Impacts of Leaks From Onshore CO2 Storage Projects on Terrestrial Ecosystems, Report 2007/03.

68. IEA-GHG (IEA Greenhouse Gas R&D Program), 2009. Long term integrity of CO2 storage – well abandonment, Report 2009/08.

69. IEA-GHG (IEA Greenhouse Gas R&D Program), 2010a. Injection strategies for CO2 storage sites, Report 2010/04.

70. IEA-GHG (IEA Greenhouse Gas R&D Program), 2010b. Pressurization and brine displacement issues for deep saline formation CO2 storage, Report 2010/15.

71. IEA-GHG (IEA Greenhouse Gas R&D Program), 2011a.

Caprock systems for CO2 geological storage, Report 2011/01.

72. IEA-GHG (IEA Greenhouse Gas R&D Program), 2011b. Potential impacts on groundwater resources of CO geological storage, Report 2011/11.

73. IEA-GHG (IEA Greenhouse Gas R&D Program), 2012. Extraction of formation water from CO2 storage, Report 2012/12.

74. (IPPC) Intergovernmental Panel on Climate Change, 2005. Underground geological storage, in IPCC Special Report on Carbon Dioxide Capture and Storage. Prepared by Working Group III of the IPCC. Cambridge University Press, New York, pp. 195–276.

75. ISO (International Standard Organisation), 2009. Risk Management – Principles and Guidelines, ISO 31000:2009. (E).

76. Javadpour, F., Nicot, J.P., 2011. Enhanced CO2 storage and sequestration in deep saline aquifers by nanoparticles: commingled disposal of depleted uranium and CO2. Transport in Porous Media 89, 265–284, http://dx.doi.org/10.1007/s11242-011-9768-z.

77. Jørgensen, E., 2007. Produksjonsteknikk: for VK 1 brønnteknikk, 1, Nesbru. Vett & Viten, 261.

78. Juanes, R., MacMinn, C.W., Szulczewski, M.L., 2010. The footprint of the CO2 plume during carbon dioxide storage in saline aquifers: storage efficiency for capillary trapping at the basin scale. Transport in Porous Media 82, 19–30.

79. Khan, F.I., Husain, T., 2003. Evaluation of a petroleum hydrocarbon contaminated site for natural attenuation using 'RBMNA' methodology. Environmental Modeling and Software 18, 179–194.

80. Khan, F.I., Hudain, T., Hejazi, R., 2004. An overview and analysis of site remediation technologies. Journal of Environmental Management 71, 95–122.

81. Koornneef, J., Ramírez, A., Turkenburg, W., Faaij, A., 2012. The environmental impact and risk assessment of CO2 capture,

transport and storage - An evaluation of the knowledge base. Progress in Energy and Combustion Science 38, 62–86.

82. Lakatos, I., Lakatos-Szabó, J., Tiszai, Gy., Palaásthy, Gy., Kosztin, B., Trömböczky, S., Bodola, M., Patterman-Farkas, Gy., 1999. Application of silicate-based welltreatment techniques at the hungarian oil fields. In: Paper SPE 56739 Presented at the 1999 SPE Annual Technical Conference and Exhibition, Houston, TX, USA, October 3–6.

83. Lakatos, I.J., Medic, B., Jovicic, D.V., Basic, I., Lakatos-Szabó, J., 2009. Prevention of vertical gas flow in a collapsed well using silicate/polymer/urea method. In: Paper SPE 121045 Presented at the SPE International Symposium on Oilfield Chemistry. The Woodlands, TX, 20–22 April.

84. Lakatos, I., Lakatos-Szabó, J., Szentes, G., Vágó, A., 2012. Improvement of silicate well treatment methods by nanoparticle fillers. In: Paper SPE 155550 Presented at the SPE International Oilfield Nanotechnology Conference. Noordwijk, The Netherlands, June 12–14.

85. Le Guénan, T., Rohmer, J., 2011. Corrective measures based on pressure control strategies for CO_2 geological storage in deep aquifers. International Journal of Greenhouse Gas Control 5, 571–578.

86. Leonenko, Y., Keith, D.W., 2008. Reservoir engineering to accelerate the dissolution of CO_2 stored in aquifers. Environmental Science & Technology 42, 2742–2747.

87. Lindeberg, E., Vuillaume, J.-G., Ghaderi, A., 2009. Determination of the CO_2 storage capacity of the Utsira formation. Energy Procedia 1 (2009), 2777–2784.

88. Loizzo, M., Duguid, A., 2006. CO_2–CementInteractions: From the Lab to theWell, EPA Technical Workshop on Geosequestration: Well Construction and Mechanical Integrity Testing , Albuquerque, NM, March 14.

89. Loizzo, M., Akemu, O.A.P., Jammes, L., Desroches, J., Lombardi, S., Annunziatellis, A., 2011. Quantifying the risk of

CO2 leakage through wellbores. SPE Drilling & Completion 26 (3 (September)).

90.   Maccormick, M.E., Hall, A.D., Trower, A.M., 2011. Component Inspection and Repair using 3D Modelling Photogrammetry Technology, Offshore Europe. Society of Petroleum Engineers, Aberdeen, UK.

91.   Manceau, J.C., Rohmer, J., 2011. Analytical solution incorporating history-dependent processes for quick assessment of capillary trapping during CO2 geological storage. Transport in Porous Media 90, 721–740.

92.   Manceau, J.C., Réveillère, A., Rohmer, J., 2011. Forcing gaseous CO2 trapping as a corrective technique in the case of abnormal behavior of a deep saline aquifer storage. Energy Procedia 4, 3179–3186.

93.   McGennis, E., 2001. Subsea tree changeout with a light subsea-well-intervention vessel, OCT 12945. In: Offshore Technology Conference, Houston, TX.

94.   Ménez, B., Dupraz, S., Gérard, E., et al., 2007. Impact of the deep biosphere on CO2 storage performance. Geotechnologien Science Report 9, 150–163.

95.   Mijnbouvvet, 2003. Mining Regulation. Regulation of the Staatssecretaris of Economic Affairs of 16 December 2002/nr WJZ 02063603.

96.   Mitchell, A.C., Ferris, F.G., 2005. The coprecipitation of Sr into calcite precipitates induced by bacterial ureolysis in artificial groundwater:temperature and kinetic dependence. Geochimica et Cosmochimica Acta 69 (17), 4199–4210.

97.   Mitchell, A.C., Phillips, A.J., Hilbert, R., Gerlach, R., Spangler, L.H., Cunningham, A.B., 2009. Biofilm enhanced geologic sequestration of supercritical CO2. International Journal of Greenhouse Gas Control 3, 90–91, http://dx.doi.org/10.1016/j.ijggc.2008.05.002.

98.   Mitchell, A.C., Dideriksen, K., Spangler, L.H., Cunningham, A.B., Gerlach, R., 2010.

99.   Microbially enhanced carbon capture and storage by mineral-

trapping and solubility-trapping. Environmental Science & Technology 44, 5270–5276, http://dx.doi.org/10.1021/es903270w.

100. Nasvi, M.M.C., Gamage, R.P., Jay, S., 2012. Geopolymer as well cement and the variation of its mechanical behavior with curing temperature. Greenhouse Gases: Science and Technology 2, 46–58, http://dx.doi.org/10.1002/ghg.

101. Nellemann, C., Corcoran, E.(Eds.), 2010. Dead Planet, Living Planet – Biodiversity and Ecosystem Restoration for Sustainable Development. A Rapid Response Assessment. United Nations Environment Programme.

102. Nghiem, L., Yang, C., Shrivastava, V., Kohse, B., Hassam, M., Card, C., 2009. Risk mitigation trough the optimization of residual gas and solubility trapping for CO2 storage in saline aquifers. Energy Procedia 1, 3015–3022.

103. Nicot, J.-P., 2009. A survey of oil and gas wells in the Texas Gulf Coast USA, and implications for geological sequestration of CO2. Environmental Geology 57 (7), 1625–1638.

104. Nordbotten, J.M., Celia, M.A., Bachu, S., Dahle, H.K., 2005. Semi-analytical solution for CO2 leakage through an abandoned well. Environmental Science & Technology 39, 602–611.

105. Nordbotten, J.M., Kavetski, D., Celia, M.A., Bachu, S., 2009. Model for CO2 leakage including multiple geological layers and multiple leaky wells. Environmental Science & Technology 43 (3), 743–749.

106. Offshore Operators Association, 2012. Guidelines for the Suspension and Abandonment of Wells, Well Operations Subcommittee on Permanent Well Abandonment, Issue 4, July.

107. Oldenburg, C.M., Unger, A.J.A., 2003. On leakage and seepage from geologic carbon sequestration sites: unsaturated zone attenuation. Vadose Zone Journal 2 (3), 287–296.

108. Oldenburg, C.M., Unger, A., 2004. Coupled Vadose Zone and Atmospheric SurfaceLayer

109. Transport of CO2 from Geologic Carbon Sequestration Sites. Lawrence Berkeley National Laboratory, Paper LBNL-55510.

110. Parrek, N., Jat, M.K., Jain, S.K., 2006. The utilization of brackish water, protecting the quality of the upper fresh layer in coastal aquifer. Environmentalist 26, 237–246.

111. Pawar, R.J., Watson, T.L., Gable, C.W., 2009. Numerical simulation of CO2 leakage through abandoned wells: model for an abandoned site with observed gas migration in Alberta, Canada. Energy Procedia 1, 3625–3632.

112. Poulsen, T.G., Moldruo, P., Yamaguchi, T., et al., 1999. Predicting soil–water and soil–air transport properties and their effects on soil vapour extraction effi- ciency. Groundwater Monitoring & Remediation, 61–70.

113. Qi, R., LaForce, T.C., Blunt, M.J., 2009. Design of carbon dioxide storage in aquifers. International Journal of Greenhouse Gas Control 3, 195–205.

114. Randhol, P., Valencia, K., Taghipour, A., Akervoll, I., Carlsen, I.M., 2007. Ensuring well integrity in connection with CO2 injection: SINTEF 2007.91.

115. Ravi, K., Boesma, M., Gastebled, O., 2002. Safe and economic gas wells through cement design for life of the well. In: Paper SPE 75700 Presented at the SPE Gas Technology Symposium, Calgary, Alberta, Canada, 30 April–2 May.

116. Réveillère, A., Rohmer, J., 2011. Managing the risk of CO2 leakage from deep saline aquifer reservoirs through the creation of a hydraulic barrier. GHGT-10. Energy

117. Procedia 4, 3187–3252, http://dx.doi.org/10.1016/j.egypro.2011.02.234.

118. Réveillère, A., Rohmer, J., Manceau, J.C., 2012. Hydraulic barrier design and applicability for managing the risk of CO2 leakage from deep saline aquifers. International Journal of Greenhouse Gas Control 9, 62–71.

119. Ristow, P., Foster, J., Ketterrings, Q., 2010. Lime guidelines for fields crops. In: Tutorial Workbook. Department of Animal Science. Cornell University, Ithaca, NY, pp. 47, Downloadable

from: http://nmsp.cals.cornell.edu/projects/curriculum.html

120. Robinson, B.A., 2010. Occurrence and attempted mitigation of carbon dioxide in a home constructed on reclaimed coal-mine spoil, Pike County, Indiana. U.S. Geological Survey Scientific Investigations Report 2010-5157., pp. 21 p.

121. Rohmer, J., Bouc, O., Fabriol, H., 2009. La question de la réversibilité dans un système évolutif, cas du stockage géologique du CO2. In: Dans Colloque interdisciplinaire réversibilité – Actes Numériques du colloque du 17 au 19 juin 2009, (in French).

122. Rohmer, J., De Lary, L., Blanc, C., Guerin, V., Coftier, A., Hube, D., et al., 2010. Managing the risks in the vadose zone associated with the leakage of CO2 from a deep geological storage. In: CONSOIL 2010 – 11th International UFZ- Deltares/TNO Conference on Management of Soil, Groundwater and Sediment, 2010, Salzburg, Austria, 22–24 September 2010.

123. Rossen, W.R., 1996. Foams in enhanced oil recovery. In: Prud'homme, R.K., Khan, S.A. (Eds.), Foams – Theory, Measurement, and Applications. Marcel Dekker, New York, NY, pp. 413–464 (Chapter 11).

124. Schramm, L.L., 1994. Foam sensitivity to crude oil in porous media. In: Schramm, L.L. (Ed.), Foams: Fundamentals and Applications in the Petroleum Industry. Advances in Chemistry, Series 242. American Chemical Society, Washington, DC, pp. 165–197 (Chapter 4).

125. ScottishPower CCS Consortium, 2011. Corrective Measures Plan – UK Carbon Capture and Storage Demonstration Competition. UKCCS – KT-S7.20-Shell- 001, Available at: http://www.decc.gov.uk/media/viewfile.ashx?filetype=4&filepath=11/ccs/sp-chapter3/ukccs-kt-s7.20-shell-001-cmp.pdf

126. Sheng, J.J., 2011. Modern Chemical Enhanced Oil Recovery – Theory and Practice. Elsevier, Burlington, MA.

127. Singh, H., Hosseini, S.A., Javadpour, F., 2012. Enhanced CO2 storage in deep saline aquifers by nanoparticles:

numerical simulation results. In: Paper 156983 Presented at the SPE International Oilfield Nanotechnology Conference, Noordwijk, The Netherlands, June 12–14.

128. Skinner, L., 2003. CO2 blowouts: an emerging problem. World Oil 224 (1).

129. Skrettingland, K., Giske, N.H., Johnsen, J.H., Stavland, A., 2012. Snorre in-depth water diversion using sodium silicate – single well injection pilot. In: Paper SPE 154004 Presented at the Eighteenth SPE Improved Oil Recovery Symposium, Tulsa, OK, USA, 14–18 April.

130. Sloan, A., Ramsay, D., Maccormick, M.E., Hall, A.D., Trower, A.M., 2011. 3D modelling photogrammetry wellhead inspection and repair. In: Paper SPE 145512 Presented at the Offshore Europe, Aberdeen, UK, 6–8 September.

131. Standard Norge, 2004. NORSOK D-010WellIntegrity in Drilling andWell Operations. http://www.standard. no/en/sectors/Petroleum/NORSOK-Standard-Categories/ D-Drilling/D-0102 (revised 03.08.04).

132. Stavland, A., Jonsbråten, H.C., Vikane, O., Skrettingland, K., Fischer, H., 2011a. Indepth water diversion using sodium silicate – preparation for single well field pilot on snorre. In: Paper Presented at the 16th European Symposium on Improved Oil Recovery, Cambridge, UK, 12–14 April.

133. Stavland, A., Jonsbråten, H.C., Vikane, O., Skrettingland, K., Fischer, H., 2011b. In-depth water diversion using sodium silicate on snorre – factors controlling in-depth placement. In: Paper SPE 143839 Presented at the SPE European Formation Damage Conference, Noordwijk, The Netherlands, 7–10 June, http://dx.doi.org/10.2118/143839-MS.

134. Storaune, A., Winters, W.J., 2005. Versatile expandables technology for casing repair. In: Paper SPE/IADC 92330 Presented at the SPE/IADC Drilling Conference,n Amsterdam, Netherlands, 23–25 February.

135. Sweatman, R., Marsic, S., McColpin, G., 2010. New approach and technology for CO2 flow monitoring and remediation. In:

Paper SPE 137834 Presented at the Abu Dhabi International Petroleum Exhibition and Conference, 1–4 November.

136. Sydansk, R.D., Romero-Zerón, L., 2011. Reservoir Conformance Improvement. Society of Petroleum Engineers, Richardson, TX, USA.

137. Sydansk, R.D., Xiong, Y., Al-Dhafeeri, A.M., Schrader, R.J., Seright, R.S., 2005. Characterisation of partially formed polymer gels for application to fractured production wells for water-shutoff purposes. SPE Production & Facilities 20 (3), 240–249.

138. Talabani, S., Chukwu, G., Hatzignatiou, D.G., 1993a. Unique experimental study reveals how to prevent gas migration in a cemented annulus. In: Paper SPE 26897, SPE Eastern Regional Meeting Proceedings, Pittsburgh, PA, USA, November 2–4.

139. Talabani, S., Chukwu, G., Hatzignatiou, D.G., 1993b. Gas channeling and microfractures in cemented annulus. In: Paper SPE 26068, SPE Western Regional Meeting Proceedings, Anchorage AK, USA, May 26–28.

140. Tao, Q., Bryant, S.L., 2012. Optimal control of injection/extraction wells for the surface dissolution CO_2 storage strategy. In: Paper CMTC 151370 Presented at the

141. Carbon Management Technology Conference, Orlando, FL, 7–9 February.

142. USArmyCorps of Engineers, 2002. Engineering anddesign. In: SoilVapour Extraction and Bioventing. Engineer Manual. Department of the Army, US Army Corps of Engineers Washington, 424 pp. (Manual No. 1110-1-4001).

143. US EPA (United States Environmental Protection Agency), 1996. A Citizen's Guide to Treatment Walls, EPA 642-F-96-016.

144. US EPA (United States Environmental Protection Agency), 1997. Design Guidelines for Conventional Pump-and-Treat Systems, EPA/540/S-97/504.

145. US EPA (United States Environmental Protection Agency), 1999. OSWER Directive 9200.4-17P – Use of Monitored Natural Attenuation at Superfund, RCRA Corrective Action,

and Underground Storage Tank Sites – April 21, 1999. Office of Solid Waste and Emergency Response, 41 pp.

146. US EPA (United States Environmental Protection Agency), 2001. Building Radon Out. A Step by Step Guide on How to Build Radon-Resistant Home, EAP/402-K-01- 002.

147. US EPA (United States Environmental Protection Agency), 2004. How to Evaluate Alternative Cleanup Technologies for Underground Storage Tank Sites. A Guide for Corrective Action Plan Reviewers, EPA 510-R-04-002.

148. US EPA (United States Environmental Protection Agency), 2008. Vulnerability Evaluation Framework for Geologic Sequestration of Carbon Dioxide, EPA430- R-08-009.

149. US EPA (United States Environmental Protection Agency), 2010a. Federal requirements under the underground injection control (UIC) program for carbon dioxide (CO_2) geologic sequestration (GS) wells: final rule. Federal Register 75 (237).

150. Vidic, R.D., Pohland, F.G., 1996. Technology Evaluation Report: Treatment Walls.

151. Vong, C.Q., Jaquemet, N., Picot-Colbeaux, G., Lions, J., Rohmer, J., Bouc, O., 2011. Reactive transport modeling for impact assessment of a CO_2 intrusion on trace elements mobility within fresh groundwater and its natural attenuation for potential remediation. Energy Procedia 4, 3171–3178.

152. Watson, T., Bachu, S., 2009. Evaluation of the potential for gas and CO_2 leakage along wellbores. SPE Drilling & Completion 24 (1).

153. West, J.M., Pearce, J.P., Bentham, M., 2005. Issue profile: environmental issues and the geological storage of CO_2. European Environment 15, 250–259.

154. Yu, J., An, C., Mo, D., Liu, N., Lee, R., 2012. Foam mobility control for nanoparticlestabilized supercritical CO_2 foam. In: Paper SPE 153336 Presented at the Improved Oil Recovery Symposium, Tulsa, Oklahoma, April 14–18.

155. Zhang,Y., Oldenburg, C.,Benson, S., 2004.Vadose zone remediationof carbondioxide leakage from geologic carbon dioxide sequestration sites. Vadose Zone Journal 3, 858–866.

Citations

CHAPTER 1

Laptev, A. , Basharov, M. , Farakhov, T. and Iskhakov, A. (2014) Calculating Efficiency of Separation of Aerosol Particles from Gases in Packed Apparatuses. Advances in Chemical Engineering and Science, 4, 143-148. Doi:10.4236/aces.2014.42017.

CHAPTER 2

A. Ghanem, J. Maalouly, R. Saad, D. Salameh and C. Saliba, "Safety of Lebanese Bottled Waters: VOCs Analysis and Migration

Studies,"American Journal of Analytical Chemistry, Vol. 4 No. 4, 2013, pp. 176-189. Doi:10.4236/ajac.2013.44023.

CHAPTER 3

Salihu Lukman, Alaadin Bukhari, Muhammad H. Al-Malack, Nuhu D. Mu'azu, and Mohammed H. Essa, "Geochemical Modeling of Trivalent Chromium Migration in Saline-Sodic Soil during Lasagna Process: Impact on Soil Physicochemical Properties," The Scientific World Journal, vol. 2014, Article ID 272794, 20 pages, 2014. doi:10.1155/2014/272794.

CHAPTER 4

He Fang, Li Haibin, and Zhao Zengli, "Advancements in Development of Chemical-Looping Combustion: A Review," International Journal of Chemical Engineering, vol. 2009, Article ID 710515, 16 pages, 2009. doi:10.1155/2009/710515

CHAPTER 5

A. Annunziatellis, S.E. Beaubien, S. Bigi, G. Ciotoli, M. Coltella, S. Lombardi, Gas migration along fault systems and through the vadose zone in the Latera caldera (central Italy): Implications for CO_2 geological storage, International Journal of Greenhouse Gas Control, Volume 2, Issue 3, July 2008, Pages 353-372, ISSN 1750-5836, http://dx.doi.org/10.1016/j.ijggc.2008.02.003.

CHAPTER 6

J.-C. Manceau, D.G. Hatzignatiou, L. de Lary, N.B. Jensen, A. Réveillère, Mitigation and remediation technologies and practices in case of undesired migration of CO_2 from a geological storage unit—Current status, International Journal of Greenhouse Gas Control, Volume 22, March 2014, Pages 272-290, ISSN 1750-5836, http://dx.doi.org/10.1016/j.ijggc.2014.01.007.

Index